NEW TRENDS IN THE HISTORY OF SCIENCE

AF745929

NEW TRENDS IN THE HISTORY OF SCIENCE

Proceedings of a conference held
at the University of Utrecht

Edited by
R.P.W. Visser, H.J.M. Bos, L.C. Palm
and H.A.M. Snelders

Rodopi

Amsterdam — Atlanta GA
1989

UNIV DUNELM

- 4 MAY 1993

509
NEW

5(09)
NEW
AVM 587

CIP-GEGEVENS KONINKLIJKE BIBLIOTHEEK, DEN HAAG

New

New trends in the history of science : Proceedings of a conference held at the University of Utrecht / (eds. R.P.W. Visser ... et al.). — Amsterdam : Rodopi. — (Nieuwe Nederlandse bijdragen tot de geschiedenis der geneeskunde en der natuurwetenschappen, ISSN 0168-9827 ; no. 30)
ISBN: 90-5183-085-8
SISO 506 UDC 5(091) (063)
Trefw.: natuurwetenschappen ; geschiedenis.

© Editions Rodopi B.V., Amsterdam — Atlanta GA
Printed in The Netherlands

CONTENTS

INTRODUCTION

The University of Utrecht was founded in 1636. Its 350th anniversary in 1986 was the occasion for a series of scientific meetings. The present volume contains the proceedings of one of these, namely the conference on "New Trends in the History of Science", which was held from 27-29 August 1986.

The conference was organised by a committee in which three groups collaborated: History of Biology, History of Mathematics and History of Science. By 1986 these groups were in the process of merging, together with the group Foundations of Physics, into one department of History and Foundations of Mathematics and Natural Science. The reorganisation intensified the contacts between the members of the groups. In this atmosphere the idea arose to organise a conference on the occasion of the university's jubilee, and thereby call attention to one of the fields around which the new department was being formed.

Within a forward-looking process, as is the creation of a new department, the discussions often turned to future developments in the field of History of Science, and in that way the subject "New Trends" suggested itself naturally as a theme for the conference.

The idea was well received by the university, by colleagues and by the scholars who were approached for a contribution to the conference. The plans materialized and the conference was held.

We look back upon the conference with pleasure and we present its results with pride. We are confident that the contributions in this volume indeed reflect significant new trends in the history of science. They deal with a wide range of subjects: science and European overseas expansion, the position of scientists under Nazi rule, the role of technology in the arms race, the emergence of the subdiscipline of solar physics, explanation in nineteenth century chemical science, a reappraisal of the role of the universities in the Scientific Revolution, an assessment of the influence of the Kantian tradition in the Romantic Era, an enquiry into the very personal scientific style of Niels Bohr, an analysis of the processes of discovery and rediscovery illustrated by the case of Mendelian heredity, and the recognition, reconstruction and testing (in part by computer simulated experiments) of a forgotten theory of polarisation.

What new trends are evidenced in these studies? This is not the place to analyse and evaluate them; but certainly there are many aspects of these contributions that are much more than novelties and have the potential for further development and application, which is implicit in the term "trend". These aspects include: the opening up of whole new areas for historical investigation, such as colonial science; the application of the methods of history of science to politically and morally very sensitive issues such as science in Nazi Germany or ballistic missile technology; the use of special methods such as quantitative analysis of scientific literature and oral history; the critical reappraisal of established pictures of scientific developments in certain periods such as the seventeenth century or the Romantic Era; and the careful analytical investigation of constituent parts of the scientific enterprise such as explanation and convention, personal style, the attribution of the status of discovery, and the disappearance, almost without trace, of viable testable physical theories.

As hosts of the conference, we invited scholars from abroad to present the new trends; at the same time we invited Dutch scholars to provide a first reaction in the form of commentaries on the main papers. These commentaries were presented at the conference immediately after the corresponding main paper; they thus served as introductions to the plenary discussions. The commentaries have been incorporated in this volume.

We express our sincere gratitude to the authors of the papers and the commentaries for their willingness to contribute to the conference, and to all others who helped to make the meeting a success.

It is a great pleasure to thank Miss Lian Hielkema and Mrs. L.M. Boersma-Broekmeyer. Their editorial assistance materially contributed to the quality of the book.

We hope that in the coming years this volume will be a source of inspiration and will challenge colleagues to take issue with the new trends. We also feel that in the longer term the contributions to the conference will serve as a valuable records, documenting trends that were new in the 1980's and inviting readers to enquire what has become of them.

Utrecht, June 1988

R.P.W. Visser
H.J.M. Bos
L.C. Palm
H.A.M. Snelders

Jed Z. Buchwald

THE INVENTION OF POLARIZATION

Almost everyone today agrees that very little, if anything, in science can be purified of theory, that, in contemporary parlance, there is no pristine language of observation.[1] Take, for example, what is generally known as Boyle's law, which asserts the proportionality of pressure to volume for gases at a given temperature. The law presumes that we know the meanings of 'pressure' and 'temperature', and so it is a theoretical statement to the extent that 'pressure' and 'temperature' are theoretical terms. And they are theoretical terms: gas 'pressure' and 'temperature', mean something quite different by the middle of the nineteenth century from what they had meant in the eighteenth.

Nevertheless even the modern skeptic tends to make distinctions based on the degree to which a given statement is theory-laden. These kinds of distinctions depend in part on general agreement about how to measure things. Pressure and temperature can be measured in ways which do not at all depend upon the particular mechanical and thermal understanding of them which obtains at a given moment. By contrast, the assertion that a given quantity of heat can be converted into a proportionate quantity of work cannot be divorced from its theoretical heritage. 'Heat' is not a thing that can be as directly measured as temperature because it is possible (in caloric theory for example) for heat to disappear without ceasing to exist as heat.

I would not like to make this distinction a hard and fast one, but it is often useful, and it is moreover one which historians frequently rely upon. In optics one does not usually think that the law of refraction considered as a formula raises the same kinds of historical issues, as, say, the question of the physical nature of light. We can, after all, easily build devices which test the law directly, but we can hardly so readily

*This article is based on a considerably longer argument which will be published in 1988 in a book entitled *The Rise of the Wave Theory of Light*. I have had to draw on many different sources to formulate my argument, and it would hardly be useful to cite them here. Accordingly I provide at the end a short list of the major sources which are directly pertinent and refer to them in abbreviated format in the text.

investigate light's structure. The law of refraction, one might say, can be discovered in the usual sense of the verb 'discover' once we have a firm understanding of how to go about measuring the things which the law involves. We have to know how to create a narrow beam of light and to measure the appropriate angles. But that's all we need to know, so that there is a sense in which the law can be 'discovered' by someone who understands how to measure the right things in the right way.

But one cannot also 'discover' the nature of light in anything like the same sense. The intricate physical assumptions which must be made are not at all so closely linked to processes of measurement as is a formula like Snel's law. One might instead say that the nature of light must be invented rather than discovered, not because I wish to make an abstruse claim about the real or unreal nature of theoretical entities, but because the elements which form a theory of the latter kind are often very far removed from processes of measurement. They always lead to measurements but, often, only through a more or less lengthy chain of deductions which only ends up with something that has the observational character of a formula like Snel's.

Since 1830 or so, at least, the polarization of light has seemed to be much more like Boyle's or Snel's law than like the equivalence of work and heat or like the physical nature of light. Certainly different theories have explained and quantified polarization in different ways, but that has not seemed to alter what one might call the fact of polarization itself. Polarization was, according to this description, discovered and not invented. Recent work that I have done on the early history of the wave theory of light has convinced me that this is entirely incorrect, that even in the first detailed experiments which the French physicist Malus performed in 1809 'polarization' had to be invented. It could not have been discovered because, as I shall try to make clear, there was more than one way to interpret measurements even at what might seem to be the most directly empirical level.

The kinds of phenomena which were united under the heading of 'polarization' in the early nineteenth century were, as is well known, first described by Huygens and used by Newton as an argument against Huygens' theory of light. A beam of light from a candle or the sun, natural light, divides into two beams when it strikes a crystal of Iceland spar. The two beams follow, in general, different paths. If, after they emerge from the crystal, the two beams are passed through a second crystal, then a peculiar thing happens. It is possible to align the two crystals so that each of the beams does not again divide in two on striking the second piece of spar. In one of the two positions where this occurs each beam follows the same kind of path in the second crystal that it followed in the first. In the other position the beams reverse roles, the second beam taking the same kind of path in the second crystal that the first beam had taken in the first crystal, and *vice versa*. In intermediate positions each beam divides in two, but in varying degrees.

Until 1809 this seemed to be a significant but unique thing since light had to pass through a crystal before it exhibited the property, which was often spoken of as requiring an 'asymmetry' or even a 'polarity' about the axis of a light ray (and so as being incompatible with Huygens' light theory, which forbid that kind of asymmetry). But in that year the French physicist Etienne Louis Malus announced he had discovered that light rays reflected from glass at a certain angle will also exhibit the same property, which he soon named 'polarization' [Malus, 1809a]. This much might seem to be descriptively unexceptional, but, in fact, I have already gone past the kind of claim made in Snel's law by asserting that the phenomenon has something to do with *rays* of light — rather than only with a *beam* of light. To grasp why it is important to distinguish between ray and beam in describing polarization experiments we must begin at the most elementary level, with the only two kinds of devices which could be used to detect 'polarization' in the early nineteenth century: the doubly-refracting crystal and the reflecting mirror. That requires pursuing Malus' early work a bit further to understand what he thought he was seeing.

Soon after his first discovery Malus made another of even greater significance to him, one which seemed closely to link the effect of reflection on a beam that has already been 'polarized' with the effect of a crystal on the same beam: to wit, that a second reflection at the proper angle will reflect a portion of the beam whose size depends on the angle between the two planes of incidence, the remaining portion being refracted. At 90° between the planes, no light at all is reflected by the second mirror. This last observation seemed to Malus essentially similar to what happens in a second crystal when it is aligned in such a way as to reverse the roles of the two beams.

As devices to detect the polarization of a beam, then, the crystal and the mirror may function in closely similar ways. Each of them generally divides a polarized beam of light into two parts, the degree of the division depending on the orientation of the parts of the apparatus. One might, and Malus in fact did, think of these analyzing devices as selecting some rays in the original beam to be affected in one way, and the remaining rays to be affected in another way. Is there anything wrong, or, rather, empirically incorrect, in thinking in this way? Indeed not — if, as I shall demonstrate in a moment, our devices are limited to crystals, mirrors and the naked eye. But, if we do think this way about what our instruments are measuring, then we are also thinking about polarization in a way that is completely incompatible with the wave theory, though it is often extremely difficult to see what the incompatibility is about, even when one has the feeling that something is not quite right. It can be so difficult to see, in fact, that it was over precisely this issue that many apparently pointless and wandering arguments took place in the 1820s and the early 1830s between the proponents of the two rival optical theories.

The best way to grasp the issue is to follow Malus in his early experiments, but I have not the time to do so here. Instead I shall starkly, and somewhat crudely, describe the essence of his understanding of polarization, an understanding he shared with most of his contemporaries [Malus, 1809b and 1811a,b,c]. As Malus understood it a beam of light consists of a collection of discrete objects called rays. These objects possess individual identities and can be counted, the intensity of a beam being measured numerically by the number of rays which it contains. Unlike the individual ray, which is too weak, a beam can be seen. Every ray has an inherent asymmetry about its length. Think of a ray as a stick to which a crosspiece is nailed at right angles. Given the direction of the ray, the orientation of the crosspiece in a plane at right angles to the ray determines the ray's asymmetry.

As Malus understood the concept, 'polarization' does not apply to the individual rays in a beam but only to the beam as a collection of rays. A beam may be polarized in a certain way, but the individual rays which make it up are not themselves said to be polarized, though each has a certain asymmetry. If the asymmetries of the rays in a given beam point randomly in many directions then the beam is, in Malus' understanding, 'unpolarized'. If, on the other hand, one can group the rays in a beam into a number of sets each of whose elements shares a common asymmetry, or even if this can be done only for a certain portion of the rays in the beam, then the beam is said to be "partially polarized". If all of the rays have the same asymmetry then the beam is just 'polarized'.

This makes Malus' conception of polarization at the most elementary level impossible to reconcile with the understanding of it developed by Fresnel, and which we hold today. According to our idea, which originated with Fresnel, at a given instant and at a given point on the wave front every wave of light possesses an asymmetry which lies in the front itself. Rays of light are mathematical abstractions: they are merely the directions joining the center of the wave to the front itself. In a purely analytical sense one can link a ray to the asymmetry in the front at the point of the front which the ray contains. Accordingly in the wave theory one can say that a *ray* is polarized, if one wishes, because polarization refers to the asymmetry at a point in the front, and to each such point there corresponds only one ray. But, and this is a signal characteristic of the wave theory, the rays, being mere lines, cannot be counted. A beam of light is not a collection of discrete rays, which means that one cannot dissect it in Malus' fashion.

To bring to the surface the intricacies of this difference consider what happens when a beam of light reflects. Suppose that this beam is "partially polarized" in Malus' sense of the term. Then it might consist of three subsets of rays. Suppose that in two of the subsets the rays have a common asymmetry, but that the asymmetry of one of these two subsets differs from that of the other. Suppose further that the rays in the third subset have randomly-oriented asymmetries. Such a beam is entirely possible in Malus' understanding and could actually be produced, Malus thought, by twice reflecting initially unpolarized light, both times elsewhere than at the polarizing angle. When we ask what happens to the 'polarization' of this beam when it is again reflected, we are in effect asking how many rays remain in each of the three original sets and whether any new sets with particular asymmetries are created. We are, that is, asking how the rays are redistributed in respect to their asymmetries by the reflection.

In the wave theory we begin with a wave striking the reflecting surface at a given instant. The wave intersects the surface in a line, and at each point along that line there is a direction, in the plane of the wave itself, which specifies the wave's polarization at that point. When we ask what happens to the polarization on reflection we are asking what happens to each of these directions. The wave account, like Malus', also deals with a *beam* of light, but it treats a beam in a very different way. In the wave theory a beam must be thought of as a number of rays. Instead, it must be thought of as a very small segment of a front. So small, in fact, that every point on the segment must have the same polarization. The vector which represents in direction this polarization may also represent in magnitude the amplitude of the oscillation at every point of that small frontal segment. On reflection the amplitude's components in and perpendicular to the plane of incidence are altered to different degrees. There results a new set of components for the reflected front at the point of incidence, and a new set of components for the refracted front at the same point. Neither the reflected nor the refracted amplitudes is the same in either magnitude or direction as the original amplitude.

To calculate the intensities of the reflected and refracted beams one must carefully consider the cross-section of the refracted beam, which must be different from the common cross-sections of the incident and reflected beams, and then use the squares of the new amplitudes to compute the energy densities in each beam. The essence, then, of the wave theoretical procedure is effectively to calculate what happens to the polarization of a given segment of the front under the approximation that the segment is so small it may be considered to contain only one point, and so to be intersected by a single ray. If we are to speak at all of rays in this context we must of necessity confound them with beams. Malus by contrast could not confound the beam with the ray because the only way he could understand the intensity of a beam was arithmetically, as a count of rays. The polarization of the beam is determined by the numbers of rays with certain asymmetries, not by the asymmetry of any one ray in the collection. One might put the difference between the two conceptions this way. The wave theory calculates the change in polarization of a ray, which it confounds with a beam. Malus never claimed to be able to calculate what happens to an individual ray, which cannot be confounded with a beam, but rather concerned himself with the redistribution of rays in respect to their asymmetries.

These two conceptions of polarization are incompatible with one another even at the level of primary empirical discourse. It is not psychologically possible for someone who thinks like Malus to describe unambiguously the results of many experiments to someone who thinks like a wave theorist, and *vice versa*, as I shall show through a specific example in a moment. The vast gulf between the two concepts made it almost impossible for scientists who did not think about polarization in the same way to communicate their experimental results to one another without leaving an inevitable residuum of ambiguity that could, and often did, lead to controversies that ended in frustration and anger.

I shall hereafter refer to Malus, and those who thought like him, as 'selectionists' since they conceived of polarization as a process in which the rays in a beam are selected and have their asymmetries altered in direction. Thusfar I have shown only that selectionists held a radically different understanding of polarization from wave theorists, and I have claimed that the two groups could not communicate even observations to one another without ambiguity. There was a material reason which made it possible for the ambiguity to go nearly unnoticed, despite the controversies it gave rise to, until it at last became irrelevant. The two devices —the crystal and the mirror — which could be used to examine polarization depended critically upon the eye to judge the presence, absence or even intensity of light, and in these kinds of experiments the unaided eye has limitations.

Consider the least ambiguous situation, the one about which selectionists and wave theorists were unlikely to misconstrue one another's empirical claims: the reflection of light at the polarizing angle. There were two ways to examine the polarization of the reflected light. The reflected beam could be passed through a crystal along the perpendicular to the crystal's surface. Then turning the crystal about the normal alters the intensities of the two beams which result in it. Or the light could be reflected at the polarizing angle by a second mirror, and then the angle between the planes of incidence on the two mirrors can be altered. Both selectionists and wave theorists would agree to the following two claims. First, set the crystal so that its principal section is parallel to the plane of reflection; then only one beam (called the ordinary beam) occurs within it: the other vanishes. Second, set the plane of incidence on the second mirror perpendicular to that on the first mirror; then no light is reflected by it; all the light is refracted into it.

About this there would have been no argument between selectionists and wave theorists. The reason no problem arises here is that the light is completely thrown into one of two possible situations: either into one of the two beams within the crystal, or else into the refracted light within the second mirror. In any other configuration of crystal or mirror this no longer holds true: there are in general two beams in the crystal, or both a reflected beam from, and a refracted beam within, the mirror. In these latter cases, which very early captured Malus' interest, selectionism and the wave theory differ markedly in what each claims must occur, but neither the crystal nor the mirror can distinguish between the claims.

Again I have not the time to explain the intricate considerations which led Malus to results which work about as well as those of the wave theory. I will instead simply give the basic structure of his account [Malus, 1811c]. Consider a polarized beam of light to strike the surface of a mirror at an arbitrary angle. To Malus such a beam consists of rays all of which have the same asymmetry, which we suppose to point in some arbitrary direction. On striking the mirror the beam has its rays sorted into two new beams, one of which is reflected while the other is refracted. According to Malus neither the reflected nor the refracted beam will any longer be polarized; each will be partially polarized only. The reflection, he argued, consists of two major subsets, each of which is in itself polarized; that is, the rays in each subset have the same asymmetries. The polarization of one subset is the same as that of the incident beam; the polarization of the other lies in the plane of reflection. The latter grows larger, while the former diminishes to nothing, as the incidence approaches the angle of polarization. The refraction also consists of two subsets. One subset has the same polarization as the incident beam: the polarization of the other lies in the perpendicular to the plane of incidence.

This is completely different from the wave theory's claim. According to it both the reflection and the refraction of linearly polarized light are also linearly polarized, because the components of the incident amplitude in and normal to the plane of reflection are altered only in relative magnitude. And these different claims have, in principle, decidedly different empirical effects. According to the wave theory a crystal or mirror used to examine the reflected beam can be turned to an angle such that only one beam of light occurs within the crystal or such that no beam reflects from the mirror. According to Malus' account this cannot be possible because the reflected beam consists of two sub-beams, each with its own polarization: if the crystal is set to produce a single internal beam from one of these two sub-beams, then the other necessarily produces two beams.

We know the wave theoretical claim to be correct. Yet Malus and many selectionists after him found nothing amiss with his account. The reason is both instrumental and theoretical. First of all, the eye cannot easily distinguish less than about a 5% or so variation in the intensities of the light which emerges from the crystal or is reflected by the mirror because the intensity in these kinds of experiments is rather small due to the minimum of two reflections which occur in the mirror experiment, and three in the crystal experiment (counting in the latter case both internal and external reflection). Secondly, the change in the angle of polarization produced by reflection and refraction is not large, amounting in absolute terms to at most two or three degrees. When we reflect a polarized beam we will, instrumentally, have only to rotate a crystal slightly out of the original plane of polarization to produce only one beam in it from the reflection. (Actually this is not, as I have stated it, true because reflection at low incidences very nearly flips the polarization to the other side of the plane of incidence, but what is true, and what Malus knew and could accomodate, is that this latter angle is very nearly the same *in magnitude* as the original angle of polarization).

Suppose, following Malus, that we have in the reflection two sub-beams, one polarized as before, and the other, much smaller, polarized in the plane of incidence. Then when we turn the crystal while observing the reflection we will, according to Malus, find a unique position for it, but a few degrees removed from the original polarization, in which one of the two beams within it is very large, while the other will be so small as to be nearly invisible. And this does indeed occur, except that, as wave theorists, we think that the "nearly invisible" beam doesn't exist at all. But, given only these kinds of experiments, Malus' extra beam could indeed exist. Precisely because the eye is relatively insensitive in these circumstances, and because reflection does not in fact deviate polarization very much, Malus' selectionist description makes qualitative sense, since a small deviation can be interpreted as the production in the reflection of a new, though small, beam with polarization in a new plane, which Malus assumed to be the plane of incidence.

Malus actually thought that he could see this weak beam, remarking that it only 'nearly' disappears. And no doubt he was correct: one can hardly so perfectly align the crystal, and so perfectly polarize the beam which enters it, that no visible light whatsoever will appear. But what seemed to Malus to be something required by theory was for Fresnel an unimportant consequence of inherent experimental limitations. When, in 1816, Fresnel realized that Malus and Biot thought otherwise he was deeply perplexed by what seemed to him their ignorance of a simple experimental fact. He never fully grasped their understanding of polarization, so he was unaware of how they could accomodate experiment.

Fresnel developed formulae for the ratios of reflected and refracted to incident intensities by projecting the incident light into components in and normal to the plane of reflection [e.g. Fresnel, 1823]. In doing so he simply assumed that the sole effect of reflection is to alter the magnitudes of the components, which at once means that the polarization will only be rotated. But the expressions for the projections which result from this procedure could not themselves be directly tested because no accurate photometric techniques were available. The only thing that could be measured, and Fresnel did measure it, was the angle of rotation, which is effectively what Malus also measured, though, unlike Fresnel, he did not calculate it. Here we find another difference between selectionist and wave theory which prevents the two from coming to terms even about experimental results, as I shall presently demonstrate.

The fact that where a selectionist insisted on seeing a small but essential light beam the wave theorist saw an inessential experimental complication would alone have been sufficient to produce an impasse between the two since both could explain the broad features of polarization experiments. But this was not all; far from it. For what has, to my knowledge, been entirely forgotten since sometime in the late 1820s is that this selectionist account of polarization leads quite directly to formulae which allow one actually to compute what will happen when polarized light reflects. In selectionist terms this means that one can compute the relative numbers of rays contained by the four beams (two reflected and two refracted) into which the original polarized light is divided. The fact that one can do so is, however, hardly obvious from Malus' memoirs or else it would certainly not have so easily been forgotten after about 1830. I discovered Malus' formulae somewhat by accident while I was studying his work, not expecting to find in it anything more than the details of his experiments and comments by him on the support they provide for the emission theory.

I would like to spend a few moments explaining how I was able to reconstruct the selectionist theory of partial reflection because this is not the only instance of its kind in the history of physics. In my experience there are at least two things which make it difficult to recognize the existence of what are by now completely forgotten theories with testable consequences. First, precisely because the theory is so thoroughly forgotten it can be exceedingly difficult to pick up the clues which point to its existence. The problem is the more pronounced in that the kind of theory I am concerned with here often leads to sentences which at first sound like what can be said using a later theory. Usually the first clue to the presence of something strange appears when it becomes impossible to reconcile with one another different statements made by the same scientist, or even different scientists, despite the fact that there may be no other indication of a change in theory. In Malus' case it gradually became clear to me that he conceived of polarization in a radically different way from the wave theorist as I realized that I could not reconcile different statements he made with one another using my understanding of polarization as a process in which a single direction remains single but is rotated.

The second difficulty arises because it can be very hard to recognize the occurrence of formulae which are based in theory. Until one realizes that there is a theory to be understood, one may not even know what it is that the scientist thinks can be quantified, much less what he has actually been able to quantify. Once I had begun to understand the pervasive influence of selectionist thinking in Malus' work I was prepared to recognize that a law which I had previously overlooked, thinking that it represented a purely empirical relation, in fact could be derived from theory. Even though Malus explicitly remarked that it could be found "from theory", I had previously ignored the comment because I could not see at all how the only 'theory' I thought he had, namely the emission theory, could possibly lead to the formula he gave, and he did not say how it did.

This formula related the increase in the amount of light refracted from a polarized beam when it is reflected with its polarization set at a certain angle to the plane of incidence, to the amount refracted at another particular setting. Since I had now begun to think that the law could be derived, even without the emission theory, I became very interested in how well it worked. Accordingly I first modeled the law on a machine; in effect I reconstructed Malus' early experiments. To my initial surprise the law worked extremely well for the range of refractive indexes Malus could have used. This could not have been coincidence — the law simply worked too well to have resulted from a guess. What, I thought, must have happened is that Malus had a way of deriving laws for partial reflection which, though no doubt incompatible with the wave theory's method, nevertheless was a good approximation to it in these kinds of experiments. I was then able to reconstruct Malus' derivation, and, once I had that, was able also to uncover the complete theory of partial reflection.

The end result of Malus' rather intricate but entirely straightforward analysis is this. Choose some fixed incidence. At that incidence reflect a beam of light polarized at 45° and measure the two angles — one for the reflection, the other for the refraction — to which a crystal must be rotated in order to produce within it what looks like (but in fact cannot actually be) a single beam. From these two angles parameters can be computed which completely determine the division of a beam with an arbitrary polarization at this angle of incidence.

Of course these two angles must be found empirically for every incidence, which seems to contrast somewhat unfavorably with the power of the wave theory, since it requires only one measurement given the index of refraction — namely, the incidence itself. Nevertheless selectionist theory could and did lead directly to a quantitative solution to the problem of the reflection and refraction of polarized light, indeed to the first solution ever to the question of how much light will be reflected, and how much refracted, at a given incidence. And until more sensitive instrumentation was developed selectionism and the wave theory had equal empirical power in this area. By the time such techniques did become available selectionism had long since disappeared for other reasons.

Furthermore, it is perhaps somewhat misleading to say that the wave theory is more powerful than selectionist theory in its account of partial reflection. The two theories do not even agree about what the important thing is to measure. According to the wave theory when linearly polarized light reflects the direction of polarization rotates, so that the crystal or mirror analyzer of the period, when set to produce a single beam or no reflection, at once shows the new direction. According to selectionism what was initially a common asymmetry possessed by the rays in the incident beam becomes two-fold, and the reflection contains two beams polarized in different directions, though one of them has many fewer rays than the other. This difference between the two theories means that they must understand the measurement process in very different ways.

The wave theorist can at once calculate the rotation of the original polarization and looks for it with the crystal or mirror. It has direct significance for him. This is not so for the selectionist. His formulae directly determine only the intensities of the two reflected sub-beams; to find how the crystal or mirror must be set to make a beam or a reflection 'nearly' invisible is not mathematically simple since one has to find the extremum of a certain sum containing several trigonometric terms, and the result is computationally intricate. But there was no need for Malus to *calculate* this minimum position since what he wished to find — the intensities of the reflection's constituent beams — emerges directly from the formulae.

Malus could then easily compute the intensity of either of the two beams in an analyzing crystal set in any way for an arbitrary angle of incident polarization (and for a given incidence on the mirror). That intensity results from the two beams which Malus' theory requires to exist in the reflection. We can of course compute the same thing using the wave theory, which is much simpler to do since only one beam is involved. I have done so and I find that the intensities the two theories require differ by less than 3% for most incidences and for all angles of polarization. (This means that had Malus gone to the inordinate effort of calculating the crystal rotation which the wave theory regards as the most important thing to measure he would have obtained good agreement with observation). Malus' formulae, then, work as well as Fresnel's for the kinds of experiments that could be performed at the time. But where the wave theorist would insist on measuring a certain angle, the selectionist would not bother, having been satisfied by the easily calculable fact, given the observation, that the intensity in the analyzer is nearly invisible when it ought to be. Put operationally, the wave theorist would first calculate the angle of rotation and then confirm the calculation by observing that, at that angle, there is, within the limits of error, only one beam in the crystal. The selectionist would (if he were asked to do so) instead first observe the angle at which there is 'nearly' only one beam in the crystal, insert the angle into his formulae, and then show by calculation that this is what should occur there because, at that angle, the 'nearly' invisible beam must be very weak.

As a second example of the power of selectionist theory to generate testable formulae, consider for a moment Biot's extensive work on what we now call chromatic polarization (but which was more often called at the time chromatic *de*polarization). This extremely influential phenomenon was discovered by Arago and involves the production of colors when polarized light which has passed through thin slices of certain crystals cut in certain ways is then either passed through an analyzing crystal or else is reflected at the polarizing angle from a mirror. Biot, Arago and most everyone else except Young and Fresnel assumed implicitly that in this phenomenon the asymmetries of the rays in the original polarized beam are selectively rotated depending upon the color which corresponds to the particular ray. Biot in effect built a quantitative selectionist theory, similar in its fundamental structure to Malus', which yielded formulae that worked rather well, in the same way that Malus' formulae worked: in determining the intensity in an analyzing crystal under given conditions [Biot, 1811 and 1812]. Biot's formulae were designed also to compute the colors involved, and here he encountered several difficulties, but these problems only made it difficult to deduce formulae entirely from selectionist principles, they did not undercut the entire endeavor.

The essential point to grasp about Biot's and Malus' work, about all selectionist analyses, is that, though theories which are based upon selectionist principles are completely irreconcilable with theories based on wave principles, nevertheless the incompatibility has very little to do with the specific *physical* principles which were usually assumed to underlie selectionism by its proponents. The conflict does not, that is, have much to do with the emission theory proper. A selectionist 'ray' does not have to consist of particles, nor refraction and reflection be explained by forces, in order to be treated as a thing: it might, for example, consist of a stream of fluid. It needs only to be something that can be counted. That assumption alone makes it possible to obtain empirically-testable consequences that work quite well given the instrumentation of the period.

Many, perhaps most, of the arguments between proponents and opponents of the wave theory from 1815 to the early 1830s can, and I believe must, be understood primarily as arguments between people who believed that rays were things and people who believed that rays were mathematical constructs. These arguments had very little directly to do with the emission theory of light, though I would not for a moment deny that almost every selectionist until the late 1820s was utterly convinced that light did consist of particles.

There are two aspects to the distinction between selectionism and the emission theory which must be remarked. First of all, there is the historical fact that not a single one of the quantitative results achieved by selectionists, and there were quite a few, depended upon the emission theory, but they did depend immediately on counting rays. Second, and even more revealing, it was possible to think that one was a theoretical agnostic as regards the emission and wave hypotheses, and yet continue to use selectionist principles, which are incompatible with wave principles.

David Brewster provides a particularly interesting example of this. Around 1830 he wrote several articles on elliptic polarization and on the structure of partially polarized and natural light [Brewster, 1830a and

1830b]. Brewster claims in them not to need any hypothesis about the nature of light. He tries, for example in discussing elliptic polarization, to make use of formulae and parameters derived from the wave theory in a way which avoids the theory proper. The result in this case is almost complete nonsense because he creates a hodgepodge which, despite his claim to have avoided theory, mixes the tacit assumption that rays can be counted with concepts that have little meaning outside of the wave theory proper. In treating polarization he seems to be entirely unaware that there is anything at all hypothetical about counting rays.

Brewster, I think, never understood the most fundamental requirement of the wave theory, that rays cannot be treated as things. Nor was he alone. Neither Biot nor Poisson ever grasped this point, and even Arago, who early promoted Fresnel's work, had difficulties understanding it. But there is also another side to this coin. It was apparently very hard for wave theorists to admit that selectionists needed only to count rays. Fresnel and Biot fought strongly and frustratingly over the issue of whether Fresnel's formulae for chromatic polarization were much the same as Biot's [Fresnel, 1821 and Biot, 1821]. I say frustratingly because Fresnel never admitted that Biot's work depended exclusively on the counting and chromatic selection of rays, while Biot did not see that Fresnel's work was incompatible with ray counting. Accordingly Fresnel attempted to attack Biot for not having properly used emission principles, which of course Biot had not since he did not need them, and Biot reacted in, I am convinced, honest and rather angry puzzlement. Neither he nor Fresnel was able to pinpoint the major issue; if it were once brought into the open, then they would have realized that their most basic concepts were so different as to prohibit almost any kind of discussion at all.

Although the difference between selectionism and the wave theory was most pronounced in polarization, it was also significant for diffraction. It has recently been argued by Naum Kipnis [1984] that, until the early 1830s, almost everyone, including those who were competent to understand the mathematics involved, thought of diffraction entirely as a process in which rays influence one another. There were two major ways to think like this. The least sophisticated simply ignored Huygens' principle and reverted to some variant of Fresnel's and Young's original method, in which only two rays are compared with one another for their path differences. The more sophisticated (empirically as well as mathematically) considered Huygens' principle to be an extension of this procedure, to be, that is, solely a method for dealing with more than two rays. Both alternatives are, in my sense of the term, selectionist because both implicitly assume that the only proper subject of optics is the ray of light.

But diffraction differs markedly from polarization in the sorts of difficulties which the conflict between selectionist and wave principles lead to. In polarization the conflict is so thoroughgoing that it could, and did, lead to arguments even about the results of experiments which in retrospect seem to be entirely straightforward. This did not occur with diffraction. Those who could follow Fresnel's mathematics (and they were few in comparison to the number who occasionally mentioned the principle of interference) could also appreciate the accuracy of his experiments, and to my knowledge no one, whether mathematically adept or not, ever did question Fresnel's experiments or interpret them in ways which were incompatible with Fresnel's formulae.

This was possible because the essential difference between selectionism and the wave theory — the reality of the physical ray — makes no difference at all in diffraction formulae, whereas it makes a great deal of difference in polarization formulae. One can consider Huygens' secondary wavelets to be loci at which rays, and not a front from which rays are derived mathematically, are emitted in all directions. The mathematical essence of Fresnel's diffraction theory is not in the least affected by this understanding because rays need not here be counted, as they must be in polarization according to selectionist theory. Huygens' principle could (and in fact was by Arago [1819]) be put as follows in selectionist terms: take a point source of light and draw about it a sphere which may be partially interrupted by an obstacle; to find the optical effect on any point beyond the sphere draw rays to that point from every point on the sphere and apply the principle of interference to the collection of rays. Of course in reality there won't be an indefinite number of rays to add up at the observation point, as is evidently assumed by Fresnel's integral formulae, but there are so many that what are in reality summations (according to selectionist theory) can be replaced by integrals. There is no need to assume that the sphere has any intrinsic physical significance other than as a locus from which rays may be drawn.

This blurring of the boundaries between selectionism and the wave theory greatly aided the dissemination of Fresnel's work because it made the principle of interference understandable as a newly discovered property of a well-known thing, the optical ray. Had this not been possible I strongly doubt that, despite the empirical power of his formulae, Fresnel's work would, as it did, have so readily won a prize offered by the Paris academy in 1818. Both Laplace and Poisson were on the prize commission, and both of them thought this way about Fresnel's formulae. Indeed a deeply revealing public controversy between Poisson and Fresnel arose four years later over Huygens' principle. The arguments were rather complicated, but the essence of the conflict amounted simply to this: Poisson [1823] refused to believe that optics could so thoroughly abandon the physical ray, as Fresnel's wave theoretical understanding of his diffraction formulae demanded.

And yet Poisson early admitted that he felt Fresnel's diffraction experiments posed grave difficulties for the emission theory-Poisson, like Brewster and many others, implicity distinguished the emission theory from selectionism by, on occasion, voicing doubts about or taking an agnostic attitude towards light particles while never for a moment abandoning the physical ray. (In fact during his controversy with Fresnel Poisson invented a way to have physical rays even if light were a propagated motion.).

If I am correct then we must, I think, reevaluate our understanding of how and why the wave theory achieved almost universal dominance by the mid-1830s. One can divide explanations for the wave theory's domination into two major classes: those which assume that the main factor was the combination of the empirical power of the wave theory with the inadequacy of the alternative emission theory, and those which recur to institutional, professional and religious factors. All of these factors unquestionably played a role. Many people, like Poisson and even Brewster, were well aware of the emission theory's empirical poverty, and this did become a favorite point for wave partisans in the 1830s to remark. The British Association, founded in 1831, was a focal point for the battle between partisans of the two theories, and its structure fostered the success of the wave men, if I may call them that. Again, the need for young mathematicians, particularly in Britain in the early 1830s, to apply mathematics to nature in the making of new discoveries almost required the wave theory if the alternative was the emission theory since one could hardly compute with the latter. Further, it also seems clear, in view of William Whewell's well-known opinions, that to many, in England at least, the wave theory seemed highly amenable to contemporary religious views (though I am quite certain that, had Whewell wished to do so, he could have theologized particles as easily as waves).

All of these explanations, useful though they are, apparently assume that the alternative to the wave theory was the emission theory, and that the issues involved in choosing between the two were altogether clearcut. Particle men understood wave men and *vice versa*, but they held different 'theories'. That assumption, I have tried to show, misrepresents the period. Those we often think of as particle men not only didn't often understand wave men, but would, and did, strongly object to characterizing much of their work as requiring the emission theory, as being worthless if the latter had to be abandoned. Though both strongly held to the emission theory neither Malus nor Biot believed that their polarization formulae were in any way hypothetical. Conversely, though many wave men, Fresnel being only the first, did perceive that their opponents formulae were hypothetical, they nevertheless felt that the formulae had somehow to derive from the emission theory and therefore stood or fell with it.

The persistence of selectionist views through the early 1830s (and even beyond in several cases) means that we cannot treat the emerging dominance of the wave theory as a change from one well understood though empirically barren scheme, based on particles and forces, to another, based on waves. The institutional, professional and religious factors which we have used to explain the transition from the emission to the wave theory must instead be applied to capture the effects of the more delicate and difficult distinctions between selectionism and the wave theory, distinctions which usually remained just below the surface of scientific discourse, subtly affecting its nature and tone. David Brewster's increasing anger at being shunted to the scientific periphery cannot be fully understood without recognizing that he remained convinced that much of his work did not require the emission theory and was *ipso facto* independent of hypothesis. He was wrong about its unhypothetical character, but none of the wave theorists of the 1830s and later knew precisely why he was wrong, though they certainly knew that he was. Brewster was not simply an old partisan of an outmoded theory who was rejected by a younger, highly trained professional elite devoted to new ideas — though he was at least that. He, and many others, were by the 1830s outsiders in an even more profound sense because he simply could not understand why his work no longer compelled attention. Someone who grew to adolescence in Japan, left for America in 1910 and returned in 1980 to die would not feel any the stranger than Brewster must have felt after the mid-1830s.

I mentioned above that the kind of theoretical difference which I have tried to explore through the example of selectionism and the wave theory seems to me to be quite pervasive in the history of physics. In electromagnetism I have argued [1985] that what one thinks of as perhaps the most elementary concept — the idea of electric charge — was so profoundly different in Britain from what it was on the Continent in the last quarter of the nineteenth century that it was nearly impossible for physicists even to understand one another. This was the more difficult in that the intricate equations required by electromagnetism often masked the conceptual difficulties until, well into a deduction or a description of experiment, an impasse was reached. Although the existence of a special British conception of charge was not so entirely forgotten as was selectionism, nevertheless the scope and nature of the idea were rapidly lost after about 1905.

These two cases illustrate something that, I feel, characterizes many controversies in physics since at least the time of Newton. Often arguments between physicists seem to revolve endlessly or to go off in apparently arbitrary directions, in the end petering out with no clear resolution, until at some time it becomes clear that the argument, unresolved though it may have been, is no longer of interest to anyone except to a few people who are by then on the periphery of the physical community. When we find irresolute controversies we may also find that the ostensible, the explicitly recognized differences are not the ones which are really at issue, and that the true issues run so deep that they have precluded mutual understanding. Over time one way of thinking becomes pervasive and the other becomes, perhaps not incorrect, but certainly irrelevant. Selectionism became irrelevant well before it became incorrect, and the electron replaced the British conception of charge well before the latter became difficult to reconcile with experiment. And yet in neither case did anyone, to my knowledge, explicitly recognize what it was that had been replaced. In electromagnetism electrons did not displace only ether models; they also displaced field discontinuities. In optics waves did not only displace particles; they also displaced rays.

Though I have insisted upon this distinction nevertheless it must not be pushed to an extreme. Fresnel was no doubt overly aggressive in forcing Biot to provide explanations based on particles and forces when Biot felt he had no need to do so, that he could remain entirely with rays. However Biot and all other selectionists of the time did reify the light ray in the optical particle, for which there was no contemporary alternative. When Fresnel attacked the emission theory he was therefore attacking the only acceptable physical basis for conceiving of a light ray as a thing. If the emission theory was called into question then so was the ray, though not in quite the same way. It remained possible to believe — and both Biot and Brewster in effect took this tack — that the physical basis of the ray had simply not as yet been discovered. Logically, of course, there is nothing at all wrong with this, but psychologically it is hardly a satisfying thing to say when the empirically-fruitful rival to the ray, the wave, did seem also to have a fairly well-founded physical basis in ether dynamics. Yet the physical ray did not disappear all at once, to be replaced by the wave; for a decade or so many people had difficulty thoroughly distinguishing between the two concepts. As wave concepts were assimilated and the theory's mathematical structure understood selectionism gradually faded, except among older men like Biot, Poisson, Brewster and perhaps even Arago who did not clearly perceive what it was that had happened. By 1860 or so physicists thought of the early years of the century exclusively as a period of conflict with the emission theory; selectionism, which had never been disentangled from light particles, had vanished even as a memory. There is a touch of irony in this when one considers that light waves were successfully disentangled from the physical ether by the mid-1890s, so that almost the entire mathematical apparatus of the wave theory remained in use. Yet early proponents of the wave theory usually cited its firm physical foundation in the ether as a major argument in its favor, and the weak physics of the emission theory as an argument against it.

University of Toronto

Bibliography

Arago, F.
1819
"Rapport fait par M. Arago a l'Académie des Sciences au nom de la commission qui avait éte chargée d'examiner les mémoires envoyés au concours pour la prix de la diffraction", in Fresnel, *Oeuvres* I: 229-246. Originally printed in *Annal. de Chimie*, 11 (1819) 5.

Biot, J.B.
1811
"Sur de nouveaux rapports qui existent entre la réflexion et la polarisation de la lumière par les corps cristallisés", *Mém. de l'Inst.*, 12 (1811) 134-280.

1812
"Sur un nouveau genre d'oscillation que les molécules de la lumière éprouvent en traversant certains cristaux", *Mém. de l'Inst.*, 13 (1812) 1-371.

1821
"Remarques de M. Biot, sur un rapport lu, le 4 Juin 1821, à l'Académie des Sciences, par MM. Arago et Ampère" In: Fresnel, *Oeuvres* I, 569-590. Originally published in *Annal. de Chimie.*

Brewster, D.
1830
"On the law of partial polarization of light by reflexion", read 4 Feb 1830, *Phil. Trans.*, 120 (1830) 69-84.

1830b
"On the phenomena and laws of elliptic polarization, as exhibited in the action of metals upon light", read 22 Apr. 1830, *Phil. Trans.*, 120 (1830) 287-326.

Buchwald, J.Z.
1985
From Maxwell to Microphysics. Aspects of electromagnetic theory in the last quarter of the nineteenth century. Chicago, 1985.

Fresnel, A.J.
1821
"Note sur le calcul des teintes que la polarisation développe dans les lames cristallisées", *Ann. phys. chim.*, 17 (1821) 102-112, 167-196, 312-316; *Oeuvres*, I, 609-653.

1823
"Mémoire sur la loi des modifications que la réflexion imprime à la lumière polarisée", read 7 January 1823, *Mém. acad. sci.*, 1 (1832), 393-433 *Ann. de chimie*, 46 (1831), 225-264,*Ann. phys. chem.*, 22 (1831), 90-126; *Oeuvres*, I, 767-799.

1866-70
Oeuvres complètes. 3 vols. Edited by H. de Senarmont, E. Verdet and L. Fresnel. Paris (Imprimérie Impériale), 1866-1870.

Kipnis, N.
1984
History of the Principle of Interference, PhD. diss. Minnesota, 1984.

Malus, E.L.
1809a
"Sur une propriéte de la lumière réfléchie", *Mem. Soc. Arc.*, 2 (1809) 143-158.

1809b
"Sur une propriéte des forces répulsives qui agissent sur la lumière", *Mém. Soc. Arc.*, 2 (1809) 254-67.

1811a
"Théorie de la double réfraction", *Mém. Sav. Etr.*, 2 (1811) 303-508.

1811b
"Mémoire sur de nouveaux phénomènes d'optique", *Mém. de l'Inst.*, 11 (1811) 105-111; *Bll. Soc. Philom. Paris*, 2:5-11.

1811c
"Mémoire sur les phénomènes qui accompagnent la réflexion et la réfraction de la lumière", *Mém. de l'Inst.,* 11 (1811) 112-120; *Bll. Soc. Philom. Paris,* 2: 320-325; *Jnl. de Phys. et de Chim.,* 73: 5-11.

Poisson, S.D.
1823
Letter to Fresnel, in Fresnel, *Oeuvres*, II, 186-189.

COMMENTARY

Casper Hakfoort *On the paper of Jed Buchwald*

An author who gives his paper the provocative title "The invention of polarization", cannot avoid asking himself, at one or another moment, an important question. The question is: how are my own historical views on the invention of polarization to be classified? Am I a discoverer or an inventor? As Buchwald did not ask this question in his paper, I would like to do so in this commentary. Therefore, I will inquire whether Buchwald's contribution contains a historiographical discovery or an invention. In my view, the paper contains both: a discovery as well as an invention.

The historiographical discovery announced in Buchwald's paper is the discovery of "selectionism", a theoretical view in optics held by Malus and others. "Selectionists" are defined by Buchwald as those who "conceived of polarization as a process in which the rays in a beam are selected and have their asymmetries altered in direction."[1] Basic assumptions of selectionist theory were that rays are "things", not mathematical constructs, and that rays can be counted.[2] Buchwald argues for the existence of selectionism rather directly. As far as I can see, his argument does not entail lengthy deductions or questionable assumptions. For this reason, the finding of selectionism can count as a historiographical discovery, of course, in the same relative way that Buchwald introduces the concept of scientific discovery.

Buchwald found selectionism in the work of Malus on polarization. He used the theoretical role of selectionism in this work as the argument for polarization being invented by Malus and not discovered. I agree with him on this, but would like to point at a terminological problem with respect to the concept of invention which is not discussed in the paper. The problem is this. The concept of invention would seem to imply that questions like: who invented something, or: when was it invented, are relevant and can be answered. I think this suggestion is wrong. Invention

1. J.Z. Buchwald, "The invention of polarization", this volume, p. 8.
2. *Ibid.*, 6.

in the history of science is not a singular event at a particular time and place. In my view, invention is a gradual and, in principle, never ending process in which new concepts or theories are formed, partly remodeled, rejected or mathematically refined, etc. So for example, polarization was partly invented by Huygens and Newton, around 1700, partly by Malus and others in the nineteenth century, and so on. Though Buchwald is not explicit on this point, he seems to hold the view (at least he does not oppose this part of the standard account) that Malus, and no one else, did invent polarization in 1809, and not at any other time.

A second comment relates in a more indirect way to Buchwald's discovery of selectionism. According to Buchwald, the law of refraction, or Snel's law, should be considered as a discovery, again in a relative sense.[3] I disagree with this. In my view the law of refraction is as much an invention as the laws of polarization. As in the case of polarization, so in the case of refraction, there was a considerable difference of opinion about the observed facts. Moreover, this difference is remarkably similar in both cases.

Before Newton's seminal paper of 1672 on the heterogeneity of light, most people who accepted Snel's law, considered it to be a law about light beams with no further specification. In particular, colours at the edges of the beam when a beam was refracted through a glass prism were considered to be secondary effects which needed an extra hypothesis to explain them. In this situation, the law of refraction was considered as a law about white light, or: light in general. The index of refraction, namely the ratio of the sines of incidence and refraction, for a certain medium like glass or water, was taken to be the index pertaining to white light. In practice (for example: in prism experiments where there was a substantial separation of the colours) people took the index of the middle colour, that is green, as *the* index of refraction. However, the important point is that they did not consider this to be the index of green light, but of white light. All this changed with Newton's new theory of light and colours, published in 1672. Newton reversed the relationship between light and colours. Colours became primary and white light became a secondary effect. In Newton's view, a beam of white light was a mixture of homogeneous rays. If, in this mixture, rays of different colours were represented in the right proportions, the resulting effect was white light. Newton therefore reinterpreted the law of refraction as a law concerning coloured rays or beams.

In the standard account of this historical episode, it is put this way: Newton *discovered* the refractive dispersion (that is: the separation of

3. *Ibid.*, 4.

colours by refraction), as well as the fact that the index of refraction depends on the colour of the light ray. In the spirit of Buchwald's analysis (though this is contrary to his explicit words about the law of refraction) it should be said that Newton invented a new law of refraction, which, of course, had some conceptual and mathematical resemblance to the earlier law. Newton's contribution was part of the invention process which led to the modern law of refraction. To sum up, Newton's theory of light and colours was a distinctly theoretical component in his law of refraction and the law, therefore, had to be invented.

Moreover, Newton's theory about coloured rays and white beams of light seems to have many parallels with the selectionist theory in Malus's account of polarization. Coloured rays can be selected and counted; in the emission theory, to which both Newton and Malus subscribed, colour is connected to a certain property of an optical particle, just like polarization is in Malus's account. Therefore Newton's view on white light and coloured rays may be an early, perhaps the first, example of the selectionist view in optics. If I am right here, it becomes quite probable that Malus was influenced in his account of polarization by Newton's use of selectionism with respect to colours.

The paper by Buchwald as I said earlier, does not only contain a discovery but also an invention. In its most pronounced form — he does make some qualifications, which I will treat later — the latter is formulated as follows: "waves did not displace particles; they displaced rays."[4] I call this thesis an invention, because this new and provocative view on a famous episode in the history of optics is founded on more or less implicit assumptions and chains of reasoning. It is, relatively speaking, more theoretical than Buchwald's discovery of the existence of selectionism; therefore his thesis about the role and importance of selectionism should be classified as a historiographical invention.

I do not agree with Buchwald's thesis, mainly because I disagree with a crucial assumption behind it. In a nutshell, my argument is this: if Buchwald holds that waves did not displace particles but rays, he is almost making a category mistake, that is: he is comparing two things which are as different as chalk and cheese.

Let me explain this by looking carefully at the concepts used in the formulation of his thesis. These concepts are wave, ray and particle. A wave can be considered as a certain movement in a material medium, e.g. the light ether. In this sense of the concept, the emphasis is on the physical basis of the wave, the physical ether. However, considering the mathematical part of the wave concept, a wave is part of an exact

4. This quotation is from the first version. In the printed version (see p. 20) the passage reads "waves did not only display particles; they also displaced rays.

description of the phenomena of light, without any commitment to a physical basis. The crucial point is that, in the first half of the nineteenth century, the concept of a light wave always had both aspects, a mathematical as well as a physical aspect. No-one at that time would talk about mathematical light waves distinct from a physical ether. Contrary to this, the concept of a light ray is essentially neutral regarding the physical question whether light is an emission of particles or a wave in a medium. The concept of a light particle, on the other hand, is a purely physical concept.

After these preliminaries, we can analyse Buchwald's thesis. When he states "waves did not displace particles; waves displaced rays", he restricts the issue to an exclusively mathematical level. This restriction is correct as far as the rays are concerned, but brings in a large amount of distortion with respect to the waves. As far as the physical part of the wave concept is concerned, waves did displace particles, despite what Buchwald says to the contrary. Of course, Buchwald is partly right. On the mathematical level, the displacement of rays by waves was indeed an important and perhaps the most important change in the first half of the nineteenth century. One of the great merits of the paper is that it draws attention to this point, but what Buchwald says simply is not the whole story. In my view, waves did displace rays as well as particles. And, as I feel that most scientists of the time considered the physical issue to be the more fundamental, I would reverse the order and say: waves displaced particles while, at the same time and at a mathematical level, waves displaced rays.

Having said this, I must add immediately that Buchwald himself does qualify his thesis, so that some of my criticism is not entirely fair. Buchwald notices that the emission theory and selectionism were *in fact* closely related, though there was — and Buchwald is right in this — no *logical* necessity for this to be so. Nevertheless, my overall impression is that Buchwald considers the mathematical issue (that is: rays versus mathematical waves) to be much more important than the physical issue (particles versus physical waves). He even seems to hold that the mathematical issue is the "true" issue.[5]

It is worthwhile to consider for a moment the context of this assumption of Buchwald. The least relevant part of this context is perhaps his obvious personal fascination with advanced mathematics, computation and refined experiments, which is apparent in his published articles and his book on Maxwellian electromagnetism. Allow me to confess here that I was very surprised when I received Buchwald's

5. *Ibid.*, 20.

congress paper to see that it contained not a single formula! A more serious observation is that Buchwald seems to hold a view of physics in which mathematics and experiments are the two dominant ingredients. In this view, physical models like those for emitted particles or ether waves, are not considered as necessary for physics. They are convenient but dispensable tools to assist with the real thing, namely mathematics and experimentation.

One additional indication that Buchwald does subscribe to this view of physics, is that he does not mention in his paper an important historical fact. Polarization in its crudest and most qualitative sense, constituted a powerful argument against the longitudinal wave theory in the early nineteenth century. In the longitudinal wave theory, contrary to the later transverse wave theory, there is no physical basis for this property of light. Thomas Young, for example, did confess that he did not know how to reconcile polarization with the physical basis of the wave theory, at that time the longitudinal wave in the light ether.[6] Here, it is relevant to note that mathematics played no role in the argument. The qualitative property of polarization represented an argument in a physical discussion, where the two alternatives were particles and physical waves.

I am not claiming that Buchwald's kind of physics, namely one in which mathematics and experiments are more important than physical models, does not play a role in history. On the contrary, it has roots in Greek antiquity and can also be traced to the epoch-making *Principia* of Newton, published in 1687. A case can even be made that this kind of physics has become dominant in our century. However, analysing history with only this narrow view of physics in mind severely distorts or misrepresents the contemporary context.

In an essay first published in 1975, Thomas Kuhn made a distinction between the mathematical and experimental traditions in the development of physical science.[7] In his view the nineteenth century was a period when the two traditions, which had not made much contact before that time, became united into what we call modern physics. Kuhn's distinction between the two traditions proved its usefulness for analysing the scientific revolution of the 17th century. Also, Kuhn rightly pointed to the nineteenth century as an important period of transition; however, I believe that Kuhn's scheme contains a faulty assumption. This is the same assumption that I see in Buchwald's analysis namely that mathematics

6. T. Young, *Miscellaneous works* (ed. G. Peacock), vol. 1 (London, 1855) 248-249 (1810), see also 229 (1809).

7. T.S. Kuhn, "Mathematical versus experimental traditions in the development of physical science", in his *The essential tension* (Chicago, 1977) 31-65.

and experimentation are the essence of physical science and its history, whereas physical models or physical world pictures do not matter in the final analysis. Both Kuhn's essay and Buchwald's paper constitute a challenge for developing a more comprehensive and subtle view of nineteenth century physics. I think that two things are essential to this new view; firstly, acknowledging that physical models, or more generally, natural philosophy, were as important as mathematics and experimentation. Secondly, looking at the nineteenth century as a period in which natural philosophical considerations, experimentation and mathematics were combined or synthesized in various combinations.[8]

In conclusion, I wish to comment on the theme of this congress: new trends in the history of science. The dominant trend in the historiography of science has been a style which had many parallels with the historiography of philosophy. Duhem, Koyré and others stressed the role of basic physical principles, the primacy of natural philosophical considerations and the ties between science and philosophy. Partly as a reaction to this dominant trend, two new historiographical trends have appeared. The first one emphasizes experimentation and instruments in the development of science and is visible, for example, in the monumental work of Heilbron on the history of electricity.[9] The other new trend, of which Buchwald is a major representative, stresses the role of mathematics and computation in the history of physics. I warmly applaud the work of Heilbron, Buchwald and others, insofar as they compensate the exclusive emphasis on natural philosophy. However, the great challenge of the present time is to integrate the older historiographical trend with the two new trends.

University of Twente

8. For a discussion of Kuhn's essay, see C. Hakfoort, *Optica in de eeuw van Euler. Opvattingen over de natuur van het licht, 1700-1795* (Amsterdam, 1986) 168-181. See also C. Hakfoort, "Newton's Opticks and the incomplete Revolution", in: G. Debrock et al. eds., *Newton's scientific and philosophical legacy* (The Hague, forthcoming).

9. J.L. Heilbron, *Electricity in the 17th and 18th centuries. A study of early modern physics* (Berkeley, 1979).

Mordechai Feingold

THE UNIVERSITIES AND THE SCIENTIFIC REVOLUTION: THE CASE OF ENGLAND

In the last two or three decades of his life, Albert Einstein increasingly assumed the roles of ambassador — as well as historian — of science. In that latter capacity he had occasion to publish an article in the *American Scientist* exactly twenty-five years ago, entitled "On the Generalized Theory of Gravitation." En route, he paused for an intriguing speculation:

> "It is fascinating to muse: Would Faraday have discovered the law of electromagnetic induction if he had received a regular college education? Unencumbered by the traditional way of thinking, he felt that the introduction of the "field" as an independent element of reality helped him to coordinate the experimental facts."[1]

Einstein went on to observe the inhibiting effect of "the almost miraculous success of the Newtonian system" on generations of scientists who, by clinging to authority, managed to rid themselves of any doubts they may have had on the idea of action at a distance. Now, I choose to begin with this episode not only because it helps bridge the chronological gap that exists between the period covered in my talk and that of the rest of the lectures in this symposium, but more important, because Einstein's comment on Faraday illumines the universality of my topic. For centuries scientists have tended to claim that the only role a formal education played in their own — or others — subsequent scientific discoveries was a negative one, and if anything, a university education impedes scientific progress. The reasons for this verdict are diverse. Many of these men of science were understandably forgetful of their own early history having greatly advanced the boundaries of their respective fields, try as they may they were in no position to recognize the relevance of their formal educational pursuits to the new frontiers of knowledge they eventually opened up. However, some of the greatest scientists added sheer defiance to selective memory, and refused to acknowledge any debt

1. *Ideas and Opinions by Albert Einstein*, trans. S. Bargmann (New York, 1954) 344.

whatever, whether to a person or an institution. Quite clearly, then, age, selective memory, and even an unwillingness to share glory with others have all contributed to the stream of uncomplimentary statements that have been made about universities down through the centuries. Add to this a natural tendency to make universities synonymous with the authoritarianism that any creative scientist will insist he — at one time — pitted himself against, and the situation is ripe for derogatory statements. Historians of science, for their part, have tended to take all such statements about universities and the early education of scientists at face value. Rarely have they made any serious attempt to scrutinize the actual educational practices of these institutions. Still less have they shown any commitment to investigating the scientific period in question in its own terms, without subjecting it to later definitions of science and yardsticks of scientific "worth."

Although such a negative view of the universities is not confined to the early modern period, the universities during the scientific revolution particularly have been singled out for abusive attention. For, to return again to our opening quotation, while Einstein and others may have gone so far as to "blame" the Newtonian tradition for its inhibiting effect on the progress of science and the education of scientists in the eighteenth and nineteenth centuries, they ultimately "forgave" its rigid authoritarianism. And the reason? Put simply, the Newtonian tradition embodied those "true" features of modern science as we know it today, i.e., mathematicized physics and experimentation. In contrast, such forbearance has never been extended to the reigning authority that Newton — or for that matter Bacon or Descartes — replaced: indeed, the Aristotelian natural philosophy that allegedly dominated the universities and the minds of men who resided within their walls has been adjudged to be no more than an endless process of syllogistic wranglings and lifeless exegetics upon the writings of the master, as far removed as possible from the study of nature itself. Again, the reason for such censure owes less to the authoritarian aspect of Aristotelianism than to its failure to conform to the scientific "expectations" of later historians. And having once judged Aristotelianism incapable of "modern" scientific vitality, historians have continually turned a blind eye to the new ideas and discoveries that, in fact, were constantly being integrated into the Aristotelian corpus. Indeed, the verdict of history has been that Aristotelianism and the universities that supposedly served as its life-support system are best forgotten and the roots of modern science sought elsewhere — in the cities and courts, or among the artisans. According to Stillman Drake, the Renaissance, in contrast to the Middle Ages, witnessed the development of the new mathematics, physics and astronomy outside the academies, while "the doors of sixteenth-century universities were closed to new scientific ideas from the outside." Another historian viewed the universities as "ineffective" in the "creation and dissemination of knowledge." Instead, they tended to be "conservative and dogmatic, controlled by the official religions of the respective countries and very slow to incorporate new knowledge." "The history of this Scientific Revolution", summarized Ashby, in his turn,

> "lies almost completely outside the universities. It is true that Harvey did his best work in Padua, and Newton taught at Cambridge. But these circumstances were incidental to their discoveries. In no sense can the universities of Europe be regarded as instigators of the Scientific Revolution."[2]

The situation in England made the universities particularly vulnerable to attack. Since the "new science" blossomed relatively late in England, and since Oxford and Cambridge of the second half of the seventeenth century were believed to have taken a total shift from the preceding century, historians concluded that the English universities were responsible for England's late start. "The science of Elizabeth's reign", thundered Christopher Hill, "was the work of merchants and craftsmen, not of dons carried on in London, not in Oxford and Cambridge." He and other historians portrayed a bleak picture according to which the English universities excluded science from their curricula — or at best, allowed for only the modicum of elementary mathematics and an antiquated natural philosphy. Such statutory negligence was allegedly accompanied by a disdain for the new science, which the universities vigorously combatted and the end result of such persecution was the forced exodus of all those men interested in such science to London and elsewhere. To quote Hill again:

> "All the greatest scientists and scientific writers of the pre-revolutionary period (though sometimes educated at Oxford and Cambridge — where else could they have been educated?) were not at the universities when they did their most important work."[3]

True, a handful of historians have tried to go beyond the standard stereotypes of the early modern universities, and in so doing have

2. S. Drake, "Early Science and the Printed Book: The Spread of Science beyond the Universities", *Renaissance and Reformation* (1970) 49; M. Kline, *Mathematical Thought from Ancient to Modern Times* (New York, 1972) 397; E. Ashby, *Technology and the Academies: An Essay on Universities and the Scientific Revolution* (London, 1963).

3. Chr. Hill, *Intellectual Origins of the English Revolution* (Oxford, 1965) 15; Idem, "Puritanism, Capitalism and the Scientific Revolution" in: Ch. Webster ed., *The Intellectual Revolution of the Seventeenth Century* (London and Boston, 1974) 248.

uncovered a wealth of new evidence illustrative of the ways in which the "new science" was, indeed, incorporated into the university curricula. Yet, by-and-large, such work has remained very much captive to older images and perceptions and is too often characterized by a placating — almost apologetic — tone.

By accepting the same frames of reference and criteria of selection, and by addressing only those issues deemed consequential by the critics, this new generation could not help but adopt, to a large extent, the language and priorities of past historians. Consequently, even in this newest work there is very little attempt to reformulate the boundaries of science to be investigated, to view early science within its larger social and cultural milieu, and, in general, to go beyond a preoccupation with those men who made "positive" contributions to science as we know it today. Hence, while this work is full of interesting material, it lacks the framework necessary for a full appreciation of the uniqueness and complexity of the early modern period. To take just the most recent example. John Gascoigne, elaborating on the work of other scholars, pointed out that the universities "could and did adopt their curricula to take account of new systems of natural philosophy." However, still seeking to accommodate older views, he hastened to add that "of course they only slowly and reluctantly abandoned the order and pedagogical convenience of their traditional studies." True, Gasgoigne observes, mathematics, mechanics, the bio-medical sciences were introduced, yet he ultimately concludes that:

> "The universities' tendency to rely on book-learning rather than experimentation in its teaching of natural philosophy is an indication that they saw their role as assisting (to cite Newman's *Idea of a University)* in 'the diffusion and extension of knowledge, rather than [its] advancement.' "[4]

I shall return later to this issue of "research", but at this point I should like to confront the most pervasive of all anachronisms to have guided historians in their evaluation of the scientific revolution in general and of the universities in particular. The anachronism concerning the relationship between "old" and "new" science, or for that matter between science and the other intellectual disciplines.

Nullius in Verba. This brief motto of the Royal Society perhaps best incapsulates the origins of the view concerning the incompatibility

4. J. Gascoigne, "The Universities and the Scientific Revolution: The Case of Newton and Restoration Cambridge", *History of Science* 23 (1985) 424.

between the "new science" and the traditional institutions of higher learning. This motto — which literally expressed the spurning of words and their manipulation in order to justify preconceived theories in natural philosophy — came to symbolize the Society's rejection of all authority and bookish learning reigning supreme at Oxford and Cambridge. And the early defenders of the Royal Society made this abhorrence of what they termed "dogmatism" the very focus of their crusade. Abraham Cowley contributed a heroic poem to Sprat's *History of the Royal Society*, in the course of which he boasted how Bacon, by destroying that "Scar-crow Deitie" of Aristotelian authority, had freed the "orchard" of philosophy: "From Words, which are but Pictures of the Thought... To Things, the Minds right Object, he is brought."[5] Joseph Glanvill went further:

> "Were I to be Judge... I think I should silence whole *Libraries* of Authors, and reduce the world of Books into a sardle: whereas were another to sit *Censor*, it may be all those I had spared, would be condemn'd to darkness."[6]

These and similar statements, when added to the previously mentioned autobiographical statements of scientists, complete the cycle of readily available material most frequently mobilized to support the bleak picture of the early modern English universities, the alleged custodians of both bookish and Aristotelian learning. And once again historians have accepted all such statements at face value, failing to grasp their polemical nature. As I have claimed elsewhere, whatever opposition Oxford and Cambridge may have expressed to the Royal Society was based, not on hostility to the new science, but rather on the traditional fear of the English universities of any attempt to establish an institution that might jeopardize their monopoly over higher education. The new institutions, for their part, were caught in an especially difficult balancing act. As much as the Royal Society — and other early academies — may have wished to quell such fears and not antagonize the universities (and we have much evidence to this effect), they had little choice but to take an aggressive stance; how else were they to attract the patronage so essential to their future success if not by differentiating between themselves and the existing institutions of higher learning. Clearly then, to attract support and money, the Royal Society had to demonstrate that it would cater to studies not already provided by the universities. There emerged, therefore, an anomalous situation whereby the supporters of the Royal Society claimed — in one breath — that they cherished the universities

RS vs Universities

5. Th. Sprat, *The History of the Royal-Society of London* (London, 1667), sig B2.
6. J. Glanvill, *The Vanity of Dogmatising* (London, 1661), "The Preface."

and intended no threat to the education they offered, and — in the next — that "real" knowledge could be pursued only within the scheme advocated and practised by the Society; such a scheme, they insisted, was not only unavailable at Oxford and Cambridge, but incompatible with the ideal of education that informed the universities.[7]

Therefore, it should be stressed that this insistent hammering on the theme of differences was largely a result of the obsession of those early polemicists with patronage and their eagerness to assure the high and the mighty of the rewards awaiting the supporters of the new ventures in natural philosophy. Thus, as Michael Hunter has pointed out, one of the most committed early members of the Society, John Beale, was totally convinced that "the spread of interest in the new science and related topics depended on vigorous propaganda, on what he called 'the right management of fame.' " In fact, the pictorial frontispiece to Sprat's *History* was originally commissioned by Beale from John Evelyn — for a similar project that Beale himself had contemplated but abandoned upon hearing of the imminent publication of Sprat's work. One cannot fail to appreciate the emblematic message depicted in the frontispiece "the prominent winged figure of Fame with her characteristic trumpet, who crowns Charles II with a laurel wreath."[8]

Not only have historians overlooked the polemical purposes of these early apologists of the Royal Society, but they've also failed to perceive that contemporary men of science by-and-large failed to conform, in their daily routine, to such exclusionist propaganda. Rare indeed was the person who practised only science or forsook philosophical speculations or did not indulge in other forms of scholarship. With few exceptions, the men of science of the second half of the seventeenth century seriously engaged in other intellectual domains, just as the general scholars — including those who attacked the Royal Society — actively pursued the new science. Not surprisingly, then, there arose voices of protest against the militant tones of the polemicists. Thus, in 1669 John Wallis reprimanded Henry Oldenburg, Secretary of the Royal Society, who seems to have been carried away with such anti-university propaganda:

> "I would not have you insist on that argument... that the University doth not meddle with Experimentall Philosophy. For it is a great mistake, (Experimentall Philosophy being as properly appertaining to the Constitution as any other...) You should rather say it is no disparagement to the Universities, for others to pursue philosophicall studies allso."

7. I have dealt with this issue in a lecture delivered at a joint session of the American Historical Association and the History of Science Society, Chicago, December 1984.
8. M. Hunter, *Science and Society in Restoration England* (Cambridge, 1981) 194-97.

And two years later, the Savilian Professor of Astronomy, Edward Bernard, similarly protested that

> "books and experiments do well together, but separately they betray an imperfection, for the illiterate is anticipated unwittingly by the labours of the ancients, and the man of authors deceived by story instead of science."[9]

It is hoped that these quotations illustrate my point. When we — as historians of science — have sought to investigate the science of the past, we have tended to forget that until the last century it was meaningless to talk exclusively in terms of that highly specialized and properly compartmentalized enterprise we recognize today as science. And by this I do not only mean that when we discuss the astronomy of the Renaissance we should also include astrology and chronology, or that alchemy should be viewed as part of chemistry. Far more important is the need to recognize that science was an integral part of a wider intellectual milieu; that despite the beginnings of the process of separation into "two cultures" — science and the humanities — very few people in our period could be considered as exclusively "mathematicians" or "astronomers" or even "physicians." These men engaged in a far wider intellectual discourse then we today care to realize; even those who displayed a visible preference for science neither exhibited a disdain for other forms of scholarship nor considered them inferior. My own studies of the writings, correspondences and library catalogues of these men attest to their continuous pursuit of both domains. And in this context we must realize that this dual engagement was not a one-way street involving only men of science. There existed a large, industrious community of "general scholars" who claimed the right to cultivate scientific studies as part of a wider range of interests and did so with great competence. Unfortunately, for all too long we have either totally neglected these men or dismissed them as — at best — antiquarians.[10]

At this point the question becomes "How can we know about these general scholars?" Not being what we today consider "scientists" they obviously did not discover or invent anything that contributed positively to the growth of science. At most their published work was of little scientific interest and in many instances they failed to publish anything at all. And the truth is, we will find information about these people only if we take the trouble to painstakingly trace those names mentioned in the

9. A.R. & M.B. Hall eds., *The Correspondence of Henry Oldenburg*, 13 vols. (Madison, 1965-73; London, 1975-86), 500; S.J. Rigaud ed., *Correspondence of Scientific Men of the Seventeenth Century*, 2 vols. (Oxford, 1841) I, 158.

10. See my *The Mathematicians' Apprenticeship: Science, Universities and Society in England, 1560-1640* (Cambridge, 1984).

correspondences of the "scientists" through the vast — though largely neglected — manuscript resources. For all too long the only manuscripts we've cared to read were those of the familiar figures about whom we knew so much already. And for this reason the examination of manuscripts of persons who, on the face of things, had little to do with science, is yet to take place. Again, my own examination of manuscripts and letters of these general scholars — regardless of whether they produced a scientific publication — reveals that they were considered respected and contributing members of the scientific community. In fact, elsewhere I've shown that many of these scholars served as important transmitters and disseminators of scientific knowledge as well as collaborators to some of their more distinguished contemporaries.

To conclude, then: If we approach the early modern period with ready-made prescriptions of who to look at and what to look for, of simplistic distinctions between reactionary universities and enlightened academies, or between ignorant dons and an independent body of scientists outside the universities, we run the risk of ignoring a large integral part of the scientific community. The result will be, not just the "regrettable but necessary" neglect of the many third-rate "inconsequentials" — the argument of those who warn we must not confuse the forest and the trees — but more important, the inevitable distortion of the lives and ideas of those very scientists so frequently researched in the first place. In fact, it is owing to this type of discriminatory scholarship that there arose the ridiculous notion that such men as Newton grew up in a vacuum, total autodidacts, neither understood nor appreciated by their contemporaries, owing nothing to either books, tutors or colleagues. A good example of such off-hand dismissiveness is the story — deriving most probably from Newton himself — concerning his undergraduate days. One day, we are told, Newton was invited by his tutor Pulleyn to join a group of students for a tutorial on Kepler's optics; the young Newton, true to form, came well prepared and consequently put everybody else to shame. Now, most Newtonian scholars have dismissed this story as apocryphal, not even bothering to make the most superficial investigation of what was taught at Cambridge in the 1660s; instead, their arbitrary, prejudiced opinions on the state of universities led them to assume that it was inconceivable for a group of seventeenth-century undergraduates to be assigned to — let alone follow — Kepler. Not surprisingly, such incredulity towards the intellectual potential of the existing educational system resulted in the ensuing unverified assumption that very few Cambridge fellows of the last decade of the seventeenth century had the vaguest idea of what the

Principia was all about. No one even mentions undergraduates.[11]

If my point concerning the need to reevaluate our perception of the mature "general scholar" is valid, it becomes doubly important to recognize the function and aims of the early modern universities. Historians ignore the fact that the universities — no matter how they may have encouraged the sciences — were not as yet expected to allow concentration on the sciences or become professional, fully autonomous, "science centers." By the late sixteenth century the English universities had ceased to fulfil one medieval function — that of preparing mainly members of the lower echelons of society for the professions, particularly for the church — and were instead expected to cater to the needs and wishes of increasing numbers of the upper classes who sought a well-rounded education. Hence, the beginnings of "secular" education for all members of the university. However, while such an education included the sciences, the concept of the autonomous status of any discipline did not as yet exist. And this brings us to the medieval concept of the unity of knowledge that survived in the universities well into the seventeenth century and beyond. At the core of such a concept was the coherence of the curriculum: the expectation that every educated man should be grounded in the entire arts and sciences curriculum, and be able to contribute to any one of its components. Therefore, we should not take the universities to task for failing to comply with our anachronistic expectations. Instead, what is extraordinary is the ease and swiftness with which they were able to accommodate the rapid assimilation of, not only the new science from very early on, but such other new disciplines as history and oriental studies as well.

Obviously, in order to suggest aspects of the topics in need of revision, I've run the risk of superficiality, for it is not within the compass of a lecture to present the necessary evidence for such claims. However, in the second half of my talk I would like to tackle one major topic in greater depth — namely, the role played by the universities as centers of scientific research. As I mentioned earlier, most historians have disputed that Oxford and Cambridge taught the new science, let alone contributed to its advancement. In order to present the evidence that contradicts such a view, I should like to approach the issue via three related perspectives. First, by defining what "research" means in the context of the seventeenth century, always bearing in mind the exact chronology of the scientific revolution; second, by clarifying the "official" attitude of the universities toward science with reference to some of the peculiarities of

11. R.S. Westfall, *Never at Rest. A Biography of Isaac Newton* (Cambridge, 1980) 83 n. 51.

the English system of higher education; and finally, by specifying some of the sources that indicate the nature of the actual "research" that was taking place within the universities.

Let us begin with chronology. As I mentioned earlier, England seems to have entered the era of the scientific revolution only during the second half of the seventeenth century. Despite the pioneering work of Gilbert, Bacon and Harvey, only after 1650 can we notice a marked shift toward the new science. Such a line of argument has proved itself a perfect ally of the compelling "Puritan thesis". According to such a reconstruction of events, the ascendancy of the Puritans to power created those conditions favourable for the cultivation of the new science; as a result of the purge of reactionary elements in the universities and the subsequent intrusion of "enlightened" Puritans, there arose those famous scientific clubs of the 1650s that eventually blossomed into the Royal Society. The more committed proponents of this thesis go on to argue that it was the Puritan "ethos" of reform — in education and in the quality of life — and emphasis on utility that was responsible for the absorption of the new Baconian, experimental science. Now, whether implicitly or explicitly, this line of argument suggests that before the late 1640s the universities opposed the new science — i.e observational astronomy, mechanical philosophy, experimental and Baconian science. What very few historians care to mention in this context, however, is that before the 1620s at the earliest, there is little sense in talking about such noble "scientific" terms. In fact, if by "new science" we mean experimental science, the mechanical philosophy and the like, it is difficult to talk about such a phenomenon before 1640. Before this date, the main scientific contributions were in the domains of mathematics, observational astronomy and optics — which, I would claim, were not at all neglected at Oxford and Cambridge — and only the beginnings of the experimental tradition, in so far as it is expounded in Gilbert's *De magnete*, were available to contemporaries for discussion. Add to this the fact that Harvey's *De motu cordis* was published in 1628; that Galileo's major works were published only in the 1630s; that Descartes's first publications appeared only in 1637, three years after Van Helmont's first publication; that even Bacon's major works did not see publication until the 1620s; and you begin to see why we cannot expect anybody — least of all an educational institution — to adopt new learning before it was invented, or at least published. Indeed, we can talk about the "mechanical philosophy" or "experimental science" only after the 1640s and 1650s, when the ideas of Descartes and Gassendi, of Boyle, Huygens and the Italian Galileans, were conducted and published. Thus, to my mind at least, university fellows and teachers were rather quick in becoming familiar with the new science both before and after 1640. As I see it, the era of the Puritan Revolution merely coincided with the spate of new publications and subsequent dissemination of major new scientific ideas. For significantly, within the universities there was just as much willingness to absorb such ideas in teaching and research before 1640 as there was afterwards. In other words, just as it would be premature to expect to find scientific societies before the establishment of the increasingly more formal meetings in Paris, London and Oxford during the 1640s and 1650s, so one must be constantly reminded about chronology in other instances. For example, the establishment of formal observatories and "open" laboratories can only be said to have become "common" in the latter half of the seventeenth century; and certainly such developments carried with them very exciting prospects for further scientific investigation. However, these new developments should not delineate our perspectives on the nature of science in previous generations. If we allow such distortion to take place, we shall not only succumb to anachronism, but miss the full significance of the earlier stages of the scientific revolution.

The second aspect demanding attention is the "official" attitude of the universities toward science in general, and the "new science" in particular. Historians have tended to define such an "attitude" on the basis of cursory readings of the university statutes, which, indeed, made little reference to science (or for that matter to any other discipline). However, the statutes in our period cannot be taken to represent curricular requirements. The main purpose of the successive codices of statutes was to regulate matters of religion and discipline, not to restrict the scope of teaching. Indeed, the statutes left the content and manner of studies largely to the discretion of the teachers, merely specifying the disciplines to be studied by each class and the most rudimentary suggestions for texts. The implicit assumption was that each lecturer would mould his course according to his own methods and incorporate additional knowledge whenever and wherever appropriate. And from the survival of a large number of students' and teachers' notebooks, it is clear that such a method of "new wine in old bottles" worked well. In fact, the statutes of the various new chairs that were found in the early seventeenth century make such an attitude explicit. The list of authorities specified is short and merely suggestive; and the lecturer is exhorted to select his own teaching methods and incorporate future developments in his field.[12]

Once we appreciate the latitude given by the universities to the

12. For this — and related points — see my paper "Laudian Statutes and Savilian Professorships" in: G. Russell ed., *The Interest of the Natural Philosophers in 'Arabick' in Seventeenth Century England* (forthcoming).

lecturers — so far as such freedom did not impinge on matters of religion and discipline — we can turn to another problematic aspect of scientific life at the universities. Oxford and Cambridge were unlike most other European universities as regards both organization and internal structure. The university was the sum total of numerous self-sufficient colleges, which not only assumed an increasing share of the university teaching responsibilities, but catered to most other needs of its members as well. In fact, one's affiliation with the university — with the exception of professorships — was dependent upon membership in a college. College fellowships, however, were limited, and most had conditions attached to them; i.e., were designated for specific counties; demanded celibacy; were restricted in length; or prohibited the occupant from receiving additional, external emoluments. Most important, however, almost without exception a college fellowship demanded the taking of holy orders and making divinity one's calling. Since the number of independent university professorships was very small, many men of science — either unable or unwilling to comply with one or more of these conditions — were forced to leave Oxford and Cambridge. But it must be stressed that the departure of these men is no reflection of any hostility of the universities to science. Instead, these men were prisoners of the peculiar patronage system in England, which acquired neither the form of the courtly tradition nor the apparatus of state patronage that were in evidence on the Continent. As regards the universities, as bodies they were incapable of supporting out of their own meagre financial resources either professors or the expensive apparatus that the "new science" demanded; instead, any additions to the existing facilities were conditional on finding private benefactors. But as we shall shortly see, the university officials did not cease to seek out potential benefactors who might be persuaded to endow the institutions with such facilities.

The reverse side of the assumption that the universities engaged solely in book learning is the assumption that they ignored the "hard core" of the scientific revolution: the use of instruments and experiments for the purpose of gathering natural knowledge and testing theories. In fact, Bennett has recently argued that the genesis of the mechanical philosophy itself owed much to the practical mathematical sciences carried on by the London mathematical practitioners, for whom the instrument was a trademark and that the practical method was a precursor of experimental philosophy.[13] Implicit in the arguments of historians who stress the importance of the mathematical practitioners is that the universities had

13. J.A. Bennett, "The Mechanics' Philosophy and the Mechanical Philosophy", *History of Science* 23 (1986) 1-28.

little sympathy for either the "base" practical sciences or the instruments used by such practitioners. The following story, related by John Aubrey, has often been used to illustrate such a contemptuous, elitist view. According to Aubrey, when Sir Henry Savile founded his chairs,

> "he first sent for Mr [Edmund] Gunter, from London, to have been his Professor of Geometrie: so he came and brought with him his sector and quadrant, and fell to resolving of triangles and doeing a great many fine things. Said the grave knight, 'Doe you call this reading of Geometrie This is shewing of tricks, man!' and so dismisst him with scorne, and sent for [Henry] Briggs..."[14]

What these historians who dismiss both Savile and the entire university on account of this story fail to realize is that Savile, not only had no objection to instruments and their employment in teaching — or for that matter, to applied science — but made their use an explicit prescription in his statutes.

> "It will, besides, be the business of the Geometry professor, at his own time... to teach and expound arithmetic of all kinds, both speculative and practical; land surveying, or practical geometry; canonics or music, and mechanics. And in explaining the above departments, I leave with the professor a free choice of the books which he chooses to explain... also, that at convenient seasons... he shall show the practice of geometry to his auditors (who choose to attend him) in the fields or spots adjacent to the university."

The professor of astronomy, for his part, was required by Savile to also teach "optics, gnomonics, geography, and the rules of navigation in so far as they are dependent on mathematics." He was further instructed "to take astronomical observations as well by night as by day, (making choice of proper instruments prepared for that purpose...)"[15]

Since it was clearly not Savile's aversion to instruments and practical mathematics that provoked this outburst (in fact, Savile bequeathed his own collection of instruments to the Savilian professors), he must have received a false impression of Gunter. He may have thought the latter's interest in instruments obsessive, perhaps even his grounding in the theoretical parts of the mathematics shaky, and all that he was capable of was toying with instruments. For significantly, another of Savile's statutory stipulations was that his professors have a solid background both in mathematics and philosophy as a precondition for election; according to his conceptions, only such a well-rounded person would be capable of the multiple tasks he had projected for his professors. Clearly, then, Gunter seems to have made a tactical error in presenting his

14. J. Aubrey, *Brief Lives*, A. Clark ed., 2 vols. (Oxford, 1898) II, 215.

15. *Oxford University Statutes*, trans. G.R.M. Ward, 2 vols. (London, 1851) I, 272-74, 277.

potential benefactor with only one facet of his talents. Significantly, Savile's view was precisely the one that, decades later, would animate the serious opponents of the Royal Society. Many of its promoters intentionally used instruments and experiments in order to draw wide support for the Royal Society and the new philosophy. Robert Boyle, for example, thus observed;

> "there are many ingenious persons, especially among the nobility and gentry, who, having been first drawn to like this new way of philosophy by the light of some experiments, which for their novelty or prettiness they were much pleased with, or for their strangeness they admired, have afterwards delighted themselves to make or see variety of experiments, without having ever had the opportunity to be instructed in the rudiments of fundamental notions of that philosophy whose pleasing or amazing productions have enamoured them of it."[16]

In the face of such outright exploitation of "gadgets" — and the inherent promise that such gadgetry would make their users proficient in the sciences — these opponents had solid ground to fear, not only philistinism, but a mortal blow to solid learning. But this is a topic for another lecture.

As for the use of instruments in the universities, a diligent examination of the sources indicates that even in the second half of the sixteenth century, university men appreciated, possessed, and used these instruments as well as were attentive to their practical applications. Many of the foremost members of the London Community were not only educated at Oxford and Cambridge, but while at university acquired those mechanical, practical, talents that would later serve them well with prospective patrons. Such an acquaintance with instruments and applied science was conveyed even to undergraduates, as is made evident by the previously quoted Savilian statutes. But there are also other sources of information. In 1592, for example, the mathematical lecturer at Cambridge, Oliver Green, received over a 60% salary increase (the annual salary of the lecturer was £ 4) "in regard to the extraordinary charges of his mathematical instruments" which he employed in his teaching. Green was modest. Three years later, the two mathematical lecturers received the additional staggering sum of £ 14 13s 4p for expenses incurred for mathematical instruments.[17] So certainly, the instruction in, and use of, quadrants, astrolabes, sun-dials and the like was not unfamiliar at Cambridge. A different type of evidence derives from the inventories of

16. M.A. Stewart ed., *Selected Philosophical Papers of Robert Boyle*, (Manchester, 1979) 4.

17. Cambridge University Archives, University Accounts 2(1), fols. 314, 325.

deceased university members. For example, we find that Andrew Perne, Master of Peterhouse, Cambridge — who died in 1580 — left behind, not only a spectacular library that included all the classics of renaissance science, but also a large collection of instruments, including quadrants, astrolabes, dials and globes. A smaller collection of instruments was left behind by Nicholas Bond, President of Magdalen College, Oxford.[18]

We also have evidence for detailed astronomical observations conducted by members of both universities, which certainly necessitated the use of proper instruments. Thus, the acquaintance of Gabriel Harvey with the people, the writings, and the instruments of the London "mathematical practitioners" was far from unique.[19] It was quite natural for such men as Wright, Briggs, Digges, Hood, Harriot, Gunter and others to move with ease between the theoretical and applied domains. In fact, the point that most historians who attach so much importance to the London practitioners fail to appreciate is the extent to which the "academic" men of science were willing to combine theory and practice, to enjoin books and instruments — *res* and *verba* — in their wide-ranging studies. On the other hand, only a handful of the "practitioners" were even remotely interested in anything that went beyond the boundary lines of their respective professions.

By the early seventeenth century, our information becomes greater. Owing to limitations of time, I would like to present only two instances of the growth of observational astronomy and experimental laboratories. For the same reason, I shall confine my examples to Oxford.

When the new Schools were opened in 1614 adjacent to the Bodleian Library, a room at the top of the tower was designated as an observatory. And following the foundation of the Savilian professorships, its occupants took charge of the room.[20] Although Savile himself began the collection of instruments for the use of his professors, it was left for the first occupant of the professorship of astronomy, John Bainbridge, to make Oxford a major center for astronomical studies. From the chance survival of Bainbridge's notebooks, we learn not only how diligent he was in the teaching of the most up-to-date astronomical discoveries, but how seriously he took upon himself the task of contributing to the growth of that science; recorded in these notebooks are both Bainbridge's on-going observations as well as his efforts to mobilize students and colleagues in

18. Feingold, *The Mathematicians' Apprenticeship*, 59, 56.

19. G.C. Moore Smith ed., *Gabriel Harvey's Marginalia* (Stratford-upon-Avon, 1913) 211-12.

20. I.G. Philip, "The Building of the Schools Quadrangle", *Oxoniensia* 13 (1948) 39-48.

cooperative observations all over the globe. In fact, it was Bainbridge who initiated that venture that was carried out by his pupil, John Greaves, in the Middle East during the late 1630s; it was Bainbridge who had arranged for an even earlier astronomical expedition to Guiana; and similarly, much of the work of Henry Gellibrand in London during the 1630s also owed a great deal to the teaching and urging of Bainbridge. These intense astronomical activities were enchanced by a rich collection of astronomical instruments including quadrants, sextants, astrolabes and telescopes. The collection, by the way, was inherited by John Greaves — who added to it — and was eventually bequeathed in 1659 by his brother Nicholas ("in memory" of Bainbridge and Greaves) for the use of the Savilian professors. Clearly then, Oxford had an uninterrupted tradition of serious, professional observational and theoretical astronomy, which in turn had replaced the more private ventures of previous decades. Thus Seth Ward, Christopher Wren and their friends during the 1650s and 1660s did not invent, but merely elaborated upon, an already existing flourishing tradition.

Similarly, the proliferation of laboratories at Oxford after the 1650s must be seen as a natural outgrowth of earlier developments. As I mentioned earlier, although the experimental tradition developed only in the second half of the seventeenth century, Oxford did not lag far behind in the earlier stages of the scientific revolution. As early as 1596 John Delaber constructed the first known laboratory at Oxford which — in line with contemporary practises — was intended in part to assist in the dispensing of chemical medicines.[22] Delaber, who was also Principal of Gloucester Hall, was part of a larger community interested in chemical/alchemical studies. Another member of his Hall was the celebrated Thomas Allen, the mentor of Sir Kenelm Digby who, in turn, maintained a major laboratory at Gresham College during the 1630s.[23]

Another early laboratory was maintained by Jacob Bobart, first keeper of the Oxford Botanical Garden, which was inaugurated in 1621. Like Delaber, Bobart used the laboratory for medical chemistry, but in line with contemporary practises, much else could be — and was — carried

21. Bainbridge's manuscripts are in Trinity College Dublin, mss. 382-386. See N. Tyacke, "Science and Religion at Oxford before the Civil War" in: D. Pennington and K. Thomas eds., *Puritans and Revolutionaries* (Oxford, 1978) 78-89 *passim;* Feingold, *The Mathematicians' Apprenticeship* 143-52 and *passim.*

22. R.T. Gunther, *Early Science in Oxford*, 14 vols. (Oxford, 1923-45) XI, 337.

23. B.J. Dobbs, "Studies in the natural philosophy of Sir Kenelm Digby, part II: Digby and Alchemy", *Ambix* 20 (1973) 146-47.

out.[24] Further investigation should reveal more information about other members of the universities who either kept laboratories in their respective colleges — as was the case with John Nidd and later Isaac Newton at Trinity College, Cambridge — or used the shops and houses of local apothecaries, as would be the practise of Boyle and the rest of the Oxford Club in the 1650s and 1660s.

The last question to be answered is: even if we accept that various members of the university studied seriously the new science, is it possible to say that officially the universities did more than just passively tolerate the eccentric practises of some of its members? Is it possible to detect a policy *encouraging* the new science? If we bear in mind the financial constraints of the universities and the peculiar structure of the English educational system, I believe we can speak of even this kind of official approval. We already mentioned the allocation of an appropriate space to serve as an observatory — and this even before the foundation of the Savilian chairs. Similarly, in 1669, the university would commit itself to the support of a professor of botany with an annual income of £ 40, certainly not in accordance with the traditional expectation that others would bear the expense. Moreover, the first incumbent, Robert Morison, also received the following year "a fine green-house or conservatory to preserve tender plants and trees."[25] And in 1681 John Fell, the "strong" man of post-Restoration Oxford seriously toyed with the idea of establishing a major observatory by expropriating the then under construction Tom Tower at Christ Church for such a purpose.[26]

But perhaps the most significant indication of the university's encouragement of the new science is the rare instance of an addition to the statutes implemented following the opening of the Sheldonian Theatre in 1667. After ordering the transfer of all disputations to the newly constructed building, this new injunction stipulated

> "that in the same place mathematical, critical, and chronological problems, and in a like manner all kinds of experiments, may be propounded and discussed, and besides, philological contents...may be engaged in by all persons of whatever degree."

Furthermore, such exercises could be opted for by anyone as part of their requirement for the M.A degree "instead of the declamations which the statutes require."[27]

24. C. Brookes, "Experimental Chemistry in Oxford 1648-1700: Its Techniques, Theories and Personnel", Oxford Honour Thesis (1985) 10.
25. R.T. Gunther, *Oxford Gardens* (Oxford, 1912) 4; "Thomas Baskerville, 'Account of Oxford' ", *Oxford Historical Society, Collectanea* IV (Oxford, 1905) 191.
26. Hunter, *Science in Restoration England*, 143.
27. *Oxford University Statutes*, II, 73-74.

But the Sheldonian was not the only foundation to symbolize Oxford's attempt to keep apace with the new science. Far more important was the opening of the Ashmolean in 1683. From its inception, the "Musaeum" was intended as a scientific institution, and as such was "the first purpose-built teaching laboratory" in England. Ever since the early 1660s members of the university had been trying to establish a laboratory and consequently were always on the lookout for possible benefactors. Hence, when Elias Ashmole proposed to contribute his own personal collection — together with the Tradescant collection he had acquired — the university agreed to bear the cost of the building, in which one floor was designated as the School of Natural History, another as the Laboratory, and the third as a home for the collections. Having made such a commitment, the university officials expected that Ashmole would also endow a professorship in natural history. In anticipation, they even conferred an *ad hominem* Professor of Chemistry upon Robert Plot, the first curator, a title later extended to Plot's successor, Edward Hannes. Ashmole, however, failed to keep his promise, and the university made no additional official appointments to this professorship, although the successors of Hannes, John Freind and Richard Frewin, appropriated a de facto title and drew their salaries from fees paid by students.[28] Once again, then, it was owing in part to the meagre financial resources of Oxford — and in part to their expectations of private benefactions — that Oxford did not do more "officially" to encourage the new science. But certainly the institution viewed the assimilation of such subjects favourably.

Some visual representations also suggest the natural process by which the new learning was assimilated by the universities. At the same time that the observatory was being built in the new schools, Thomas James, first Bodley librarian, was arranging for the painting of the library's freeze. And there, among the greatest contributors to knowledge of all ages and in all disciplines, were the geographers Mercator and Ortelius and the astronomers Copernicus and Brahe.[29] Two decades later, the abridged version of the new Laudian Statutes — for the use of students — included a pictorial representation of the Copernican system; the person responsible for this version was Peter Turner, Savilian professor of geometry as well as one of the compilers of the Laudian code. The emblematic representation of the triumph of the new learning after the

28. For the Ashmolean see A.V. Simcock, *The Ashmolean Museum and Oxford Science 1683-1983* (Oxford, 1984) and references there cited.

29. J.N.L. Myers, "The painted freeze in the picture gallery", *Bodleian Library Record* (1950-51) 87.

Restorian is found again in the magnificent ceiling of the new Sheldonian; Theology, naturally, is in the center, Mosaic Law and the Gospel on each of her sides.

However,

> "opposite to *Theology*, are *Mathematick Sciences* in several squares (depending on *demonstration* as the other on *Faith*). As *Astronomy* with the *Celestial Globe*, *Geography* with the *Terrestrial, Geometry* with her *Square* and *Plumb-line, Optick* with her *Perspective-glass*; Also *Arithmetick, Perspective*, &c."

Elsewhere, we witness the

> "*Herbalist* imploring the assistance of *Truth* for the right understanding of her Plants. The *Chymist* with a *Bolt-head* in his hand. The *Chyrurgeon* as preparing her self to finish the dissecting of an head, which hath the Brain already opened is held before her..."

And; in the midst of all,

> "as descending from above, is the figure of *Truth* sitting on a Cloud in one hand a *Palm-branch*, the Emblem of Victory, in the other the *Sun*, whose Brightness enlightens all the Circle with Figures, and is so bright, that is sometimes hides the face of her self to us that are below."[31]

Certainly a worthy counter emblem to Evelyn's frontal piece to Sprat's *History of the Royal Society*. One also wonders whether Robert South, during that often quoted attack he is supposed to have waged against the new science and the Royal Society at the Sheldonian in 1669, ever bothered to look up.

Appropriately, it was in the Sheldonian theatre that Joseph Addison delivered a speech during the Act of 1693, in "Defence of the New Philosophy." He described the downfall of Aristotelian philosophy and the role of Descartes and Boyle as the chief architects of the new philosophy. He began, however, with one of the earliest public encomiums to Isaac Newton:

> "How long, gentlemen of the university, shall we slavishly tread in the steps of the ancients, and be afraid of being wiser than our ancestors. How long shall we religiously worship the triflings of antiquity, as some do old wives stories. It is indeed shameful, when we survey the great ornament of the present age, [Newton] to transfer our applausses to the ancients, and to take pains to search into ages past for persons fit for panegyric."[32]

30. Tyacke, "Science and religion at Oxford", 89-90.
31. *A Description of the Painting of the Theater in Oxford* (Oxford, 1673).
32. B. le Bovier de Fontenelle, *A Conversation on the Plurality of Worlds*, trans. W. Gardiner (London, 1777) 141.

I hope that the previous few examples have helped to illustrate my claim for the need for a major reevaluation of the role of the universities in the scientific revolution. For too long we have been captive to slogans and distorted perceptions, quick to applaud anyone who attacked "bookish learning" or criticized Aristotle. With the same indiscrimination, we've all too readily dismissed the universities from serious consideration, preferring instead to view them as reactionary, monolitic institutions resistant to any and all change. And while such misconceptions clearly thrived on the relative scarcity of documentary evidence, I hope that today, if anything, I've shown that the sources are far richer than previously believed, if only we cast a wider net. And so, let not the scarcity of evidence be interpreted as the evidence of scarcity!

Boston University

COMMENTARY

H. Floris Cohen *On the paper of Mordechai Feingold*

I enjoyed this paper tremendously. Rather than an enumeration of chronologically arranged facts on a take-it-or-leave-it basis, it presents a specific, unorthodox, and stimulating thesis in a clear-cut, forceful way. I also like very much the larger, cultural framework into which an important, although here somewhat underrepresented, segment of the history of science has been put by our speaker.

After these complimentary statements have been made, you may rightly suspect that here is the point at which the 'but', or 'however', appears. However, before I shall elaborate my fairly substantial 'but', I wish to insist on the non-perfunctory nature of my initial statement. Feingold's paper has the great merit of forcing one to rethink afresh one's own position on the issue at hand. For this we may be truly grateful, even though I also believe that our speaker's brilliant rhetoric falls short in substantiating the far-reaching claims he makes in his paper on behalf of the scientific progressiveness of the universities during the Scientific Revolution, particularly in England.

I deliberately choose here the term 'progressiveness' so as to open my position to the accusation directed so very freely and frequently by our speaker against virtually all other historians who have addressed this topic. Although he never actually uses the word 'Whiggish' history writing, this is what he aims at all the time. In Feingold's view, apparently, history of science in connection with the unversities has always in effect meant Whiggish historiography, that is, history which imposes anachronistic standards on the particular part of the past one is dealing with. I think that the speaker has far overextended the proper use of what seems to have become quite a shibboleth in history-of-science polemics. At the end of my comment I shall briefly come back to this issue.

Before I reach that point, I should like to split my critique into three parts, which deal with the following issues:

1. Feingold's claim, though on the surface of it very clear, requires some finer distinctions in order to be made really unambiguous.

2. The few empirical facts Feingold adduces serve at best to sustain a far weaker case than the one he so forcefully presents.
3. We seem to forget a little bit what the Scientific Revolution was all about.

I shall now address these three points in turn.

Under the overall umbrella of sizing up the importance of the universities in the Scientific Revolution, Feingold in my view conflates a great variety of quite distinct possible notions and conceptions. Let me list a few, in random order.

One might claim, for example, that

- the universities hampered the Scientific Revolution;
- the universities fought the Scientific Revolution;
- the universities were irrelevant to the Scientific Revolution;
- the universities failed to disseminate the Scientific Revolution;
- the universities failed to originate the Scientific Revolution,

and so on. Of course these possible positions are not necessarily mutuallly exclusive, or meant to be exhaustive. My only point is that these are different possible conceptions of the role of the universities in the process of the Scientific Revolution, but that Feingold attacks them indiscriminately. For example, he ridicules the notion that Aristotelianism was incapable of absorbing the new scientific notions which emerged in the course of the 17th century. He also castigates Ashby for simply stating that the universities failed to instigate the Scientific Revolution. Now these are statements of quite different scope. My suspicion is that, in implicitly handling these statements as if they are more or less on a par, our speaker tries to argue the stronger case by searching for evidence which applies to the weaker case only. Indeed, whereas Feingold provides at least some indication that some assimilation of some new scientific findings took indeed place at some universities, he provides no shred of evidence for a notion that the universities played a major contributing role in the process of the Scientific Revolution. I, for one, will continue to consider any such conception as a remarkably tricky one to uphold, unless and until Feingold provides evidence which squarely matches this particularly far-reaching subclaim of his.

Thus I have already started to address my second criticism. Not only is it often left unclear by Feingold what particular general point is aimed at by what piece of empirical material, I think the evidence Feingold does in fact adduce is rather meagre overall. It may well be significant that the one and only fact contained in the sorty related to us by Feingold that would really succeed in beginning to shake my admittedly traditional, unoriginal, scarcely researched, fairly negative view of the role of the universities in the Scientific Revolution - namely, Kepler's optics being taught at Cambridge in Newton's student days - is not positively affirmed by our speaker - he only ridicules the negation of the assertion. So I would like to ask: was Kepler's optics taught at Cambridge university during Newton's student days, or not, and if so, in what curricular framework? And if not, or if we don't know, what was the point of bringing the matter up to begin with?

Let me just touch upon one more example of the scantiness and lack of solidity of the empirical material presented in the paper. Feingold intriguingly suggests that part of the relative lack of science-orientedness of England's universities was due to the competition for patronage with the Royal Society as a result of the meagreness of their financial resources. By implication, in a country such as the Netherlands, where no scientific society came into existence until the middle of the 18th century, no such barrier existed; yet Dutch universities were, if anything, later than their English counterparts in picking up and more or less fruitfully assimilating some of the new scientific insights. This leads me to my third point.

Let us try not to forget entirely what the Scientific Revolution was all about. Was it about observational astronomy *per se*? No: observational astronomy had existed from Babylonian times onwards; what broke essentially new ground was rather certain specific changes in planetary theory. Or was the Scientific Revolution about the use of scientific instruments? No: in astronomy, surveying, musical theory, and other parts of applied mathematics instruments had been employed ever since the Age of Exploration at the latest. Or was the Scientific Revolution about talk of combining reason and experience? No: such talk, as distinguished from the real thing, goes back at least to Aristotle.

In sum: everything Feingold tells us about university people doing all these things is irrelevant to the issue of what the universities meant for the Scientific Revolution, as distinct from what they meant for 17th century science and culture at large. *The* Scientific Revolution is an *analytical* concept, which should not be reified. It does not cover everything in the 17th century that went under the name of science, or, more properly speaking, of natural philosophy. Rather, the concept of the Scientific Revolution refers to the mathematization of nature as pursued by men like Galileo and all the other traditional heroes; to the systematic application of experiments; to the formulation of a mechanistic world view. These are the core features of the Scientific Revolution, and scarcely a trace of them is to be found in the scattered examples of scientific interest on the university level that Feingold has laid before us.

There is no denying, or ignoring, or passing by, such significant facts as can be taken from the biographies of most people who mattered to the process of the Scientific Revolution, such as is done very effectively in ch. 6 of Westfall's undergraduate textbook *The Construction of Modern Science*. To add just one example, Christiaan Huygens studied at Leyden, but the only teacher from whom he could learn anything worthwhile at all was Frans van Schooten - a (at least among academics) little appreciated professor who chiefly taught in the vernacular at what can best be termed an affiliated polytechnic. Of course, the protagonists of the Scientific Revolution were rooted in the science of their own time, which was largely that of the university curriculum. But what makes these men stand out is precisely the extent to which they managed to transcend these origins. In this sense Feingold's paper has failed to convince me that I should give up my orthodox views on the matter under discussion.

One final point. There has been a time when accusing earlier generations of historians of Whiggishness, or of crude anachronisms in their approach to history, was a valuable and timely contribution toward a proper method for studying the history of scientific ideas. Increasingly I have come to believe that this period is over. The injunction not to be Whiggish has turned into a shibboleth: it is now almost indiscriminately being hurled at everybody who happens to believe that over the ages science has exhibited some measure of progress. True, the fact of scientific progress should not close our eyes for realizing that science has not progressed along a unilinear, predetermined path. Buth this is not now at issue in discussing Feingold's paper. All the genuine pleasure I had in reading, and in listening to, his paper does not prevent me from asserting that it far overshoots its mark in denying a number of important insights gained by an earlier generation of historians of science concerning the very minor role played by the universities in bringing about or in sustaining the Scientific Revolution.

University of Twente

Frederick Gregory

KANT'S INFLUENCE ON NATURAL SCIENTISTS IN THE GERMAN ROMANTIC PERIOD

Recent studies of the Romantic Era have come a long way from the older view that this period holds little of interest to the historian of natural science. The Romantic Period has of late emerged as a time which holds many lessons for those who seek to understand the social and intellectual forces that determine how science is viewed, how it is organized and regulated, and to what degree it is valued in society. We are coming to realize far better how the unique political, social, and intellectual frameworks of the late eighteenth and early nineteenth centuries conditioned Germans to regard natural science as they did.

This study deals primarily with the intellectual factors that contributed to the evaluation of natural science in this period. It has two major concerns: 1) an examination of the fundamental differences between two important intellectual communities from this time, the Kantians and the romantic *Naturphilosophen*, regarding the role and significance of natural science, and 2) an assessment of the ways in which the relatively little known Kantian tradition of the Romantic Period survived and continued to influence the development of natural science.

In the second half of the nineteenth century Otto Liebmann's rallying cry, "Back to Kant!", confirmed that the system of the Königsberg philosopher could be invoked as a philosophy particularly fit for an age in which natural science played an increasingly dominant role. The great physicist and physiologist Hermann von Helmholtz explicitly asserted that the most recent developments in mathematics and in the experimental investigation of sensation were consistent with Kant's teaching, and Friedrich Albert Lange, in his famous *Geschichte des Materialismus*, employed an approach sympathetic to Kant in order to correct the philosophical excesses of scientific materialism that had caught the public's attention in the 1850s.[1] Kant's thought presented a

1. Cf. my *Scientific Materialism in Nineteenth Century Germany* (Dordrecht, 1977) 33. On Liebmann and Helmholtz, see Lewis W. Beck, "Neo-Kantianism", in: *The Encyclopedia of Philosophy* (New York, 1967), Vol. V, 468-469.

sober corrective to the overly simplistic and radical extremes of materialistic metaphysics. At the same time it provided a philosophical interpretation of natural science that employed explanations consistent with principles of Newtonian mechanics, an interpretation that recognized a central role for empirical observation in the discovery and determination of causal laws. In short, a Kantian understanding of the nature and goal of natural science was very appealing to practicing scientists after mid-century.

As is well known, the neo-Kantian movement of the second half of the nineteenth century was also a response to what was perceived to be excesses of the German idealisitic philosophy of the first several decades of the century. To many natural scientists at mid-century, thinkers like Fichte and Hegel represented men who held nature in little regard when compared to the allegedly superior and determinative role of mind. In spite of their philosophical erudition, these thinkers were looked back on as dogmatists who believed that they could arrive at the truth of nature without having to rely on empirical observation. But when men such as Johannes Müller, and especially his students, included Schelling's philosophy of nature among the systems of idealistic philosophy that they rejected, they clearly wished to forget that Schelling's system *had* been different from the others, enough so that it had once been able to captivate many great scientific minds. Had the very same scientists who found Kant so appealing at mid-century asked themselves why the Königsberg philosopher had not fared as well among scientists during the *early* decades of the century, they would have had to admit that there had been something about Schelling's treatment of nature that had overshadowed Kant, and that whatever it had been, it was no longer persuasive.

In this paper I should like to examine some of the implications of both Kant's and Schelling's philosophy of science, with an eye to the question: Why was Kant eclipsed by Schelling early in the nineteenth century only to re-emerge later as the philosophical preference of many German scientists?

There can be no doubt about the impact of Kant's critiques on the German intellectual community at the end of the eighteenth and beginning of the nineteenth centuries. Philosophers quickly realized the seminal nature of Kant's monumental achievement and the immensity of the challenge it presented. But if anyone could be presumptuous enough to try to match it, it was the thinkers of *Sturm und Drang*.- "It was a matter of course in that day and region", wrote Josiah Royce, "that if you were young, and were anybody at all, you were a genius. The only question was what sort of genius in your lordly spiritual freedom you had chosen to be".[2] If you had chosen to be the philosper genius, Royce continued, your business was "first of all to transcend Kant, and secondly to transcend everybody else".[3]

Now what was it about Kant that was unsatisfactory to Schelling and his age? In the intellectually lively atmosphere of the years around 1800 several factors weakened Kant's influence on natural scientists of the Romantic Period. I should like to mention three.

In the first place Schelling and others understood Kant to have superceded the dogmatism of rationalism and the disunity of empiricism only by restricting what could be known of nature so severely that one could no longer claim one was dealing with the real. In his memorial essay on Kant Schelling said that he wished to correct Kant's "merely negative and inadequate resolution of the conflict between reality and abstraction".[4] Kant *seemed* to be saying that knowledge of *nature* was not really possible, since experience itself was only made possible by the categories employed by the *mind.* Scientists were ultimately lawgivers more than they were discoverers of law, since the *a priori* conditions governing the possibility of experience were at the same time the sources from which all general natural laws must be derived.[5] This was perceived to mean that knowledge of nature was more dependent on mind than it was on nature.

Nor was Schelling alone in this perception of Kant. Even before he discovered Schelling's *Naturphilosophie*, G.H Schubert came to similar conclusions about Kant's inability to satisfy what he found to be captivating about natural science. At the end of the eighteenth century he wrote to Emil Herder about his new-found love of science with the words "I have now found something real".[6] But Schubert did not view Kant as a friend of the natural scientist; in fact, he saw Fichte's hatred of nature as a working out of the Kantian tendency. Writing to his parents in 1798 about an essay he had composed on what he called "the Kantian-Fichtean egoism", Schubert added the terse judgment — "which I hate".[7]

2. Josiah Royce, *The Spirit of Modern Philosophy* (New York, 1967) 171.
3. *Ibid.*
4. Quoted by Norbert Guterman in N. Guterman ed., E.S. Morgan trans., *F.W.J. Schelling, On University Studies* (Athens, Ohio, 1966) 156, n. 53.
5. Cf. Kant's *Prolegomena zu einer jeden künftigen Metaphysik* in *Gesammelte Schriften* (Berlin, 1911), Vol. IV, 320.
6. Nathanael Bonwetsch ed., *G.H. Schubert in seinen Briefen. Ein Lebensbild* (Stuttgart, 1918) 34.
7. *Ibid.*, 10-11.

Shortly after this he became acquainted with Schelling, and there he found what he sought.

Schelling sought to resist the claim he thought Fichte and Kant were making, namely, that the mind was more real than nature. What Schelling preferred was a perspective in which nature was at least as real and autonomous as mind. Ludwig Feuerbach, looking back on the early Schelling from later in the century, understood that the original direction of Schelling's *Naturphilosophie* was not, as is too frequently thought among historians of science today, the same as that seen in Kant and Fichte.

> "Schelling, as is known, wanted originally to travel the opposite or reverse path from idealism. *Naturphilosophie* was in actuality at first nothing but inverted idealism, hence the transition from the latter to the former was not difficult ... Nature in *Naturphilosophie* was not something derived or fixed, but something primary and independent".[8]

Rarely has Schelling been portrayed in this light. As a result, except for those who have read and digested his works, few have been disposed to appreciate Schelling's sympathies for a sound empirical foundation to any philosophy of nature.[9]

A second factor conditioning and weakening the reception of Kant's thought also stemmed from its restricted focus. For all its apparent novelty, Kant's critical approach shared with post-Cartesian rationalism the assumption that scientific explanation should be modelled on and confined to what may loosely be called mechanical explanation. To be sure, Kant's understanding of mechanics referred to Newton's dynamical approach, not that which arose out of the so-called "mechanical philosphy" of the seventeenth century. But Schelling, because in his broader focus he was willing to attribute to nature a freedom and autonomy impossible in Kant, came to understand the meaning of nature's rationality differently from Kant. Nature for Schelling could not be dictated to or confined by Kant's categories of human knowledge; hence, nature could not be captured through mechanical explanation

8. Ludwig Feuerbach, "Kritik der Schellingschen Philosophie", in: M. Frank and G. Kurz, *Materialen zu Schellings Philosophischen Anfängen* (Frankfurt, 1975) 399-400.

9. In the *Einleitung zu dem Entwurf eines Systems der Naturphilosophie* of 1799 Schelling corrected the misunderstanding on the part of some that a philosophy of nature was a purely deductive armchair enterprise. "The assertion that natural science must be able to deduce all its principles *a priori* is in a measure understood to mean that natural science must dispense with all experience, and, without any intervention of experience, be able to spin all its principles out of itself — an affirmation so absurd that the very objections to it deserve pity". F.W.J. Schelling, *Sämtliche Werke* (Stuttgart, 1958), Vol.III, 278.

alone. Schelling appealed to a broader context of human perception than mechanical interaction, one closer to the whole of our experience as living beings. Nature was not a timeless and immutable machine, but a temporal and developing organism. All being, including God Himself, evolved. In his essay on human freedom Schelling wrote:

> "Has creation a final goal. And if so, why was it not reached at once. Why was the consumation not realized from the beginning. To these questions there is but one answer: Because God is *Life*, and not merely being... Being is sensible only in becoming".[10]

According to Arthur Lovejoy it was this insertion of radical evolutionism into Schelling's understanding of nature that allowed him to turn the age-old Platonistic scheme of the universe upside down. Nature's rationality was not to be expressed through its conformity to fixed and immutable laws of the mind; nature's rationality was the rationality of becoming, of growth, of fulfillment of a pre-ordained plan. It is hard to overstate the new turn taken in Schelling's rejection of the static conception of nature's rationality in favor of a dialectical one. What is important to recognize is that it resulted from attributing more autonomy to nature than was possible in the Kantian perspective, and that it too contributed to the reasons why Schelling's *Naturphilosophie* was assumed to be more real and more appropriate for the natural scientist than Kant's philosophy of science.

A third factor affecting Kant's influence on natural scientists has two parts. It was the confusion that reigned in the scientific community at the turn of the century over certain implications of Kant's thought for specific questions in 1) philosophy of science, and 2) in the relation between natural science and religion. While Kant had been vitally interested in natural science, his purpose had not been to focus on or even to work out in detail the implications of his system for specific questions in the philosophy of natural science or for the interaction between science and religion. Such issues may not have been unclear in the mind of the great thinker from East Prussia, but they remained undeveloped in his written works. To many who came after him, they were not clear at all. The existence of this confusion also contributed to weakening Kant's influence on those thinkers whose primary concern was the status of natural knowledge.

10. Quoted from Schelling's *Über das Wesen der Menschlichen Freiheit* of 1809 by Arthur O. Lovejoy in *The Great Chain of Being: A Study of the History of an Idea* (New York, 1960) 318. The explicit assertion of the primacy of organism and development is also evident in Schelling's earlier work. In the *Einleiting zu dem Entwurf eines Systems der Naturphilosophie* of 1799 Schelling wrote "We behold in what is called nature not the primal product itself, but its evolution", *Sämtliche Werke*, Vol. III, 299-300.

One such confusion had to do with the distinction between theory construction and metaphysics. Peter McLaughlin has shown in his recent analysis of Kant's relationship to the great anatomist Samuel Thomas Soemmerring how one of the best scientists of the time fundamentally misunderstood what he thought Kant would say was the difference between a metaphysical question and a theoretical question of natural science.

According to McLaughlin, Soemmerring thought that any part of natural science that was not purely descriptive was transcendental. As a result he sought approval from Kant for his assertion that the seat of the soul was located in the fluid of the ventricles in the brain, a question he called an *a priori* problem of "transcendental physiology" that led "into the most distant realms of metaphysics".[11] Kant had to lecture the confused scientist not only on the impossibility of locating the seat of the soul anatomically, but also on mistaking a scientific hypothesis about the location of the soul's *activity* in the body for a metaphysical question. McLaughlin sees the second error to be much more serious than the first. He notes that Kant wrote three drafts of his Afterword to Soemmerring's *Organ der Seele* of 1796, not because he was being careful in providing the definitive commentary as an expert in philosophy, but because Soemmerring had put him in the delicate position of having to instruct a practicing scientist about scientific theory.[12] Soemmerring, writes McLaughlin, had "made himself a layman in his own discipline".[13]

Another problem not addressed clearly by Kant had to do with the nature of scientific theories, how they were formed, and how their status varied from one natural science to another. How, for example, was one to move beyond the empirical data of the senses. For Kant the laws of chemistry would always be "merely empirical", since they were generalizations that were not constitutive of the mind in the way the laws of Newtonian mechanics were. And where organic phenomena were concerned, there was confusion among Kantians about what sense observation entailed. Some, like Kant himself, contended that organic purposiveness was a fact of experience, while others argued that our observation of organic nature did not direct us to apply a nonmechanical teleological principle to living things.

Schelling's *Naturphilosophie* seemed to offer hope that here finally was

11. Peter McLaughlin, "Soemmerring und Kant. Über das Organ der Seele und den Streit der Fakultäten", in: Gunter Mann and Franz Dumont eds., *Samuel Thomas Soemmerring und die Gelehrten der Goethezeit* (Stuttgart, 1985) 192.

12. *Ibid.*, 199.

13. *Ibid.*, 200.

a system which was going to clarify what remained unclear in Kant on matters like induction.[14] But Schelling proved ultimately disappointing, for he too could not solve the problem satisfactorily. If nature and mind *must* agree because they were both manifestations of the same underlying reality (as Schelling's system required), then, in spite of Schelling's claim that both inductive and deductive means could be used to arrive at that reality, induction from empirical observation no longer seemed to be free and unbiased. As Heinrich Friedrich Link put it, the *Naturphilosophen* were trying to ban doubt from science, a prospect he believed to be harmful for science.[15]

Finally there was confusion about the implications of Kant's approach concerning the relationship between science and religion. In addition to Schelling, the dominant theologian of the time, Friedrich Schleiermacher, who far preferred Schelling's philosophy of nature to Kant's, agreed that a problem with Kant's system was that it implied all consideration of nature must be removed from theology. This sharp separation of science from religion was viewed by both Schleiermacher and Schelling as another unfortunate result of Kant's "inadequate resolution of the conflict between reality and abstraction".

Recent theology has cast new light on this perception. Ueli Hasler writes that in the theology of feudal times, when nature was assumed to be totally dependent on God's will, nature was an instrument through which God could work out His purposes. Idealists such as Kant and Fichte destroyed feudal nature, according to Hasler, by erecting the mind as a new authority over it. Fichte interpreted nature as something that did not exist by itself, but was the product of the spontaneously active *Ich*. Such an interpretation removed nature from dependence on God and shifted the Deity's traditional role to human beings, thereby further emancipating humanity from transcendent religous authority.[16] In Kant, and more particularly in neo-Kantian theologians, the effect was similar, but not as radical. By denying that nature could be known through religious means, and by declaring that God could not be known in nature, Kantians too contributed to the removal of nature from nineteenth century theology. Such an implication was not at all welcome to most thinkers of the Romantic Period.

14. This was the reason why, for example, Jakob Fries was initially attracted to Schelling. Cf. my "Die Kritik von J.F. Fries an Schellings Naturphilosophie", *Sudhoffs Archiv*, 6 (1983) 153-156.

15. *Ideen zu einer Philosophischen Naturkunde* (Breslau, 1814) iii-iv.

16. Ueli Hasler, *Beherrschte Natur. Die Anpassung der Theologie an die Bürgerliche Naturauffassung im 19. Jahrhundert* (Bern, 1982) 26-40.

Seen in Marxist terms, this retreat to a de-natured theology was the predictable accompaniment of the shift from feudal to capitalistic society. Hasler asserts that the attack on the religiously legitimized feudal world view could only be successful if nature was seen as a world that was lawful to itself, a world that existed by itself, a world that was bereft of God *(entgöttert)*, in sum, a world of things and events that were empirically verifiable.[17] Kantian theology, as worked out by Jacobi and Fries in the Romantic Period, sounded the battle cry to separate natural science and religion.[18] At the beginning of the nineteenth century, however, Germany was unprepared for either a capitalistic society or for a de-natured theology. Not until after 1848, during the years of unification and the rapid economic expansion of the *Gründungsära*, would the dominant theology of the day harken back to Kant. In the neo-Kantians Albrecht Ritschl and Wilhelm Herrmann the German reaction to Darwin could be far more tolerant than that of other schools in which it was still necessary to weave the knowledge of nature into the fabric of theological cloth.

Schelling himself recognized the radical trend in Kant's approach, but many practicing natural scientists did not. They could only sense a preference for the relation between science and religion found in Schelling and Schleiermacher to that contained in Kantians like Jacobi and Fries. A complete separation of science from religion found few takers in the Romantic Period.

These three factors, then — that in Kant nature seemed somehow less real than mind, that Kant's scientific description of nature had to be restricted to mechanical interaction alone, and the confusion that reigned about the status of scientific theory and the relation of science to religion — all worked *against* Kant and for Schelling and *Naturphilosophie* at the beginning of the nineteenth century. They help us to understand why Schelling was preferred by many over Kant.

Having examined some of the factors negatively affecting Kant's reception as a philosopher of science in the Romantic Period, and having a better appreciation why Schelling was preferred by many scientists in the early years of the century, let us now inquire how Kant was influential when he *was* invoked by scientists.

In the first place it can be noted that Kant's thought, in contrast to Schelling's, was more relevant to the practical side of science. The great

17. *Ibid.*, 26.

18. The sharp separation of science from religion is especially clear in Fries's *Wissen, Glaube, und Ahndung* of 1805. Cf. J.F. Fries, *Sämtliche Schriften* (Aalen, 1968), Vol. III, 413-755.

weakness in Schelling's approach was its lack of practicality where research was concerned. In spite of his ridicule of the accusation that he ignored the need for an empirical foundation for his conclusions, it remained true that Schelling's concern was only *philosophical* in the end. He may have emphasized his respect for the empirical when he wrote that "we originally know nothing at all except through experience and by means of experience, and in this sense the whole of our knowledge consists of the data of experience", but his real concern became evident in his next sentence: "These data become *a priori* principles when we become conscious of them as necessary".[19]

The lack of practicality in Schelling's thought derived from the very characteristic that made Schelling's approach attractive to so many, for if nature was equally as autonomous as reason, the structure of reason could be exhibited as well from the real world as from the ideal. But while the *Naturphilosophen* produced impressive examples of the agreement between nature and mind, their purpose was to make a philosophical point, not to pursue practical research programs. Eventually they were even viewed by many as opposing empirical labors.

Such a fate did not befall scientists who were willing to appeal to Kant. While neo-Kantians too struggled against the notion that scientific investigation was a mere piling up of facts, even providing a definitive critique against crude empiricism,[20] they were not accused of opposing empirical research.

In another study I have argued that the concept of university education Kant threw up to the Prussian authorities in his *Streit der Fakultäten* of 1798 differed markedly from the conception of university study Schelling preached in his lectures on the method of academic study in 1803 precisely on this point.[21] Schelling declared openly that German *Wissenschaft* should *not* be taught with a practical or technical intent. As a result of this understanding of *Wissenschaft*, which prevailed in these formative years of university reform in Germany, science and technology were separated from each other in German education, not integrated as

19. Schelling, *Einleitung zu dem Entwurf*, in *Werke*, Vol. III, 278.

20. Imre Lakatos credits Jakob Fries with being the first one to emphasize that "propositions can only be derived from other propositions, they cannot be derived from facts". Cf. his "Falsification and the Methodology of Scientific Research Programs", in: I. Lakatos and Alan Musgrave eds., *Criticism and the Growth of Knowledge* (Cambridge, 1970) 99. Cf. also Karl Popper, *The Logic of Scientific Discovery* (New York, 1968) 93-94, 104f.

21. "Problems in the University Administration of Science in the German Romantic Period", paper delivered in the session "The Politics of Science" at the Annual Meeting of the History of Science Society, October, 1986.

they were in France. Not only was Schelling's rejection of the practical, experimental dimension of research not to be found in Kant, but this major difference between the two was consistent with their philosophical disagreement about the nature of intuition. Kant insisted on sense intuition, denying the possibility of intellectual intuition, while Schelling specifically abandoned sense intuition in favor of intellectual intuition. In general the Kantians supported and encouraged practical scientific research, while Schelling and the *Naturphilosophen* did not. These implications for university education must, of course, be viewed in conjunction with the larger social transformation from *Gelehrtentum* to *Bildungsbürgertum* about which Steven Turner has written so eloquently.[22]

A second way Kant's lingering influence was visible in the Romantic Period occurred whenever a justification of proper scientific explanation was required. After all was said and done in Kant's turgid critiques, what emerged as the acceptable standard for explanation in natural science was Newtonian mechanics. Kant's famous definition of "genuine science", requiring as it did that scientific explanation be cast in mathematical terms, could not have been farther removed from the goals of Schelling and his followers. "I hold", wrote Kant in the Preface to the *Metaphysische Anfangsgründe der Naturwissenschaft*, "that in every special natural theory one encounters genuine science only to the extent that one encounters mathematics in it".[23]

As a result of this uncompromising sentiment, those who continued to view mathematical physics in its historical role as the paradigm science could claim new authority for their position against others who, like Schelling and for that matter Goethe as well, would replace the mechanical approach to nature with something more lively. Kant's understanding of genuine science served as a point of departure for those who harbored an unease and impatience with Schelling and his program. Heinrich Friedrich Link, who wrote on zoology, botany, chemistry, physics, and geology during his scientific career, composed several works on the relation between natural science and philosophy in the early years

22. See the following works by Turner: "The *Bildungsbürgertum* and the Learned Professions in Prussia, 1770-1830. The Origin of a Class", *Histoire sociale – Social History* 1 (1980) 105-135; "The Growth of Professional Research in Prussia, 1818-1848 —Causes and Context", *Historical Studies in the Physical Sciences* (1971) 137-182; "University Reformers and Professional Scholarship in Germany, 1760-1806", in: Lawrence Stone ed., *The University in Society* (Princeton, 1974) 495-531; and "The Prussian Professoriate and the Research Imperative, 1790-1840", in: *Epistemological and Social Problems in the Development of the Sciences in the Early Nineteenth Century* (Dordrecht, 1981) 319-339.

23. Kant, *Gesammelte Schriften*, Vol. IV, 470.

of the nineteenth century. In the third one, entitled *Ideen zu einer philosophischen Naturkunde* and published in 1814, Link agreed that "mathematical physics ... is rightly called the fundamental science of natural knowledge".[24] Link conceded that it could tell us nothing about an inner living principle in nature. But he insisted that one must start with mathematical physics and with it come to understand its own limitations. He expressed sympathy for Schelling's goal to investigate the world soul lying beneath phenomena, but he showed considerable impatience with what he felt was Schelling's unjustified starting point.[25]

Besides Link there were numerous others who were trained by the Kantian natural scientists in Göttingen. As is well known, Kant was regarded as one of the most important philosophical minds of the day among Göttingen scientists.[26] But clearly the most articulate spokesman for the Kantian position during the Romantic Period was Jakob Friedrich Fries, who taught physics and philosophy at Heidelberg and Jena. Fries was among the first to oppose Schelling's *Naturphilosophie*, and it was Fries who was the first to criticize Goethe's *Farbenlehre* after it finally appeared in 1810.

Fries so valued Kant's definition of genuine science that he wrote an entire volume spelling out its implications. In his *Mathematische Naturphilosophie* of 1822 Fries gave himself the task of justifying and describing why the only comprehensible scientific knowledge of nature available to humans stemmed from the phenomena of motion. Motion was the sole natural phenomenon that could be constructed mathematically, and mathematical constructability, as Kant had taught, was the condition required in any genuine scientific explanation. Like Kant Fries concluded that Newton's laws of motion were of a metaphysical origin. Fries credited Kant with correcting Newton's assumption that the mathematical formulation of natural philosophy came *after* the knowledge of nature had been acquired from experience. Kant, said Fries, gave mathematics the dominant role, *prescribing* the highest laws of motion and the forms of nature's basic forces for all natural knowledge. To justify this right of mathematics was one of the major tasks he set for himself in his book.[27]

24. Link, *Ideen*, 76.
25. *Ibid.*, v, 77.
26. McLaughlin, "Soemmerring und Kant", 196. Included among Göttingen faculty and students who were known to possess respect for Kant were G. Lichtenberg, J.F. Blumenbach, C.H. Pfaff, C.F. Kielmeyer, G.R. Treviranus, J.F. Meckel, C.F. Reil, C. Girtanner, A. Scherer, and A. von Humboldt.
27. Fries, *Die Mathematische Naturphilosophie nach Philosophischer Methode Bearbeitet*, in: *Sämtliche Schriften*, Vol. XIII, 32.

Thirdly, Kant's influence was also decisive among German chemists. At least two scholars present are no doubt better informed than I about German chemistry of this period. Professor Snelders has written extensively of Kant's influence on the inorganic natural sciences in general and on J.B. Richter in particular, while Professor Hufbauer's work has shown us how German chemists came together into a community at the end of the eighteenth century.

Philosophical discussions about chemistry in the period following Kant invariably began with a consideration of the question of atomism versus dynamism. Kant exerted a decided influence here, for few if any Germans opted for the atomism of the English over Kant's dynamic theory of matter. "Atomism", wrote Fries in 1808, "explained the filling of space by solely the presence of mass in it. Kant showed this was incorrect and demonstrated that space could only be filled by the mutual limitation of attractive and repulsive forces".[28] As long as atomism remained the fundamental conceptual perspective, Fries observed fourteen years later, too many accidental characteristics of matter escaped confinement by law. But Kant freed chemistry of this inadequacy by showing that the decision between atomism and dynamism was to be made on *a priori* grounds. Only a purely mathematically determined relationship could be raised to the status of a principle, whereupon natural science reached its completion. Forces possessed this capacity, said Fries, atoms did not.[29]

Fries conceded that all experimental chemists spoke of atoms, and he agreed that they should be allowed to. But they should understand that "atoms" do not represent the simplest forms of matter, only matter that has not yet been encountered as compounds.[30] Practical compromises of this sort characterized most German chemists in the Romantic Period.[31]

From all this one can see that some romantic Kantian scientists did not accept Kant's claim that even the broadest generalizations in chemistry rested on merely empirical foundations and not on *a priori* principles of knowledge. Elsewhere I have argued that both J.B. Richter and Fries, in their attempts to mathematize chemical forces, were completely at odds with Kant about the status of chemistry. The irony was, of course, that

28. Fries, "Atomistik und Dynamik", in: *Sämtliche Schriften*, Vol. XVIII,237.

29. Fries, *Sämtliche Schriften*, Vol. XIII, 24, 28-29, 504.

30. *Ibid.*, 239.

31. Cf. Alan J. Rocke, "The Reception of Chemical Atomism in Germany", *Isis* 7 (1979) 535; H.A.M. Snelders, "Atomismus und Dynamismus im Zeitalter der Deutschen Romantischen Naturphilosophie", in: Richard Brinkman ed., *Romantik in Deutschland* (Stuttgart, 1978) 187-201.

Richter's stoichiometry may well have contributed to the development of Dalton's nondynamical atomism.[32]

The fourth and final example of Kant's influence on natural science I shall consider occurs in the life sciences. Here Kant's influence was felt in a manner similar to that he exerted in chemistry; namely, Kant inspired some to argue, against the master himself, that biology enjoyed genuine scientific status. Jakob Fries, for example, identified as one of Kant's greatest errors his assertion that the purposive organization of living things, which allegedly defied mechanical explanation, was evident in the empirical apprehension of nature itself. According to Fries it was not the case that a limitation of our faculty of understanding made a complete mechanical explanation or organism impossible. Organisms could be described with mechanical means, said Fries. A teleological judgment of organism was not forced upon us by nature but was brought to the observation by the observer.[33]

Ironically, this strict mechanical approach to organism, which would find a more sympathetic ear later in the century, came to Fries from reading Schelling's *Naturphilosophie.* Schelling impressed Fries with having freed science from Kant's mistake of separating organism from those objects of science that could be captured by characteristic laws of nature. What Fries inherited from Schelling was the conviction that the world apprehended by scientists was one organized whole; it did not consist of some phenomena that were susceptible to scientific description and others that were not. Of course where Schelling tied everything together under the rubric of organism, Fries united natural phenomena under mechanism. To do so Fries conceded that to arrive at mechanical laws more was required than merely the *a priori* considerations of mathematical natural philosophy. Fries attempted to make clear how formal conjecture and inference could be of service in uncovering general principles of natural science. In so doing he bequeathed to his student Ernst Apelt in Jena and to philosophers thereafter the problem of induction.

Many biological scientists of the time did not agree with either Schelling or Fries about the status of organism. In his description of the several stages of the teleomechanical program in nineteenth century German biology, Timothy Lenoir has shown how Kant's philosophical treatment of organism directly influenced the German community of

32. F. Gregory, "Romantic Kantianism and the End of the Newtonian Dream in Chemistry", *Archives Internationales d'Histoire des Sciences*, 3 (1984) 108-123.

33. Fries, *Sämtliche Schriften*, Vol. XVII, 245-248.

biologists throughout the Romantic Period and beyond.[34] Suffice it here to say merely that Kant's beliefs that the purposive organization encountered in living things could not be explained or understood through a strictly mechanical account of organism, that biological science must nevertheless carry mechanical explanation as far as it would possibly go, and that it should take the fact of purposive organization as an unexplained starting point all struck sympathetic cords among numerous biological scientists of the era here under review. Kant's carefully articulated view was not always comprehended; some of those who identified with his position overstepped the limits Kant himself had imposed. But there can be little doubt that the defense of teleology among practicing German biologists owed much more to Kant than it did to Schelling.

In summary, Kant's strict sense of genuine science persisted into the Romantic Period, influencing those concerned with natural science in different ways. While some preserved Kant's intent to impart first class scientific status to mathematical physics alone, ascribing a qualified and derivative rank to chemistry and biology, others insisted that all natural phenomena were, at least theoretically, subject to mathematical description. Further, because of its association with Newtonian science, Kant's strongly mechanistic philosophy of natural science not only equipped those who preferred it with a respected means of opposing Schelling's organic philosophy of nature, but it also, through its restricted but steady focus on the phenomena of experience, reinforced reverence for the empirical labors of *practicing* scientists. Eventually Schelling moved beyond his program of *Naturphilosophie* and transcendental philosophy to the more mystical level of the philosophy of identity. As *Naturphilosophie* was transformed into unabashed German idealism, the Kantian position appeared more and more viable as the appropriate philosophy for a natural scientist. The Kantian lamp was by no means extinguished during the Romantic Period; but it only began to burn brightly again after mid-century.

University of Florida

34. Timothy Lenoir, *The Strategy of Life: Teleology and Mechanics in Nineteenth Century German Biology* (Dordrecht, 1982).

COMMENTARY

H.A.M. Snelders *On the paper of Frederick Gregory*

An interesting and characteristic aspect of the history of Dutch science is that the Netherlands often served as a receptacle for new, 'foreign' scientific ideas. It is common knowledge that René Descartes's ideas were quickly accepted in the Republic, both inside and outside the universities. Isaac Newton's empirical-mathematical natural science was spread over the continent by means of publications of the triumvirate Herman Boerhaave, Willem Jacob 's Gravesande, and Petrus van Musschenbroek. The first scientists to adhere and defend Antoine Laurent Lavoisier's new chemistry were Dutch: the Haarlem physician Martinus van Marum and, in particular, the merchant/chemist Adriaan Paets van Troostwijk, his friend, the physician Jan Rudolph Deiman and their Amsterdam "Society of Dutch Chemists". Not only did they publish several excellent articles on the new system, they also carried out a great number of original experiments supporting Lavoisier's theory.

At the turn of the eighteenth century the few Dutch philosophers restricted themselves to a popular kind of philosophy with an emphasis on practical applications.[1] Well-known representatives were Frans Hemsterhuis, Daniel Wyttenbach, and Philip Willem van Heusde. Immanuel Kant's new philosophy was quickly accepted in the Netherlands.[2] The most important question was: do the fundamentals of Kant's philosophy support or do they endanger Christian belief? It was evident that Kant's "moralische Vernunftglaube" could not possibly be accepted by orthodox Christians. In particular the Amsterdam minister Paulus van Hemert defended Kant in a loyal, enthusiastic, and ardent way. He published a book: *Beginzels der Kantiaansche wijsbegeerte* (Principles of Kantian philosophy; 1796-1798) and edited a *Magazijn voor de critische*

1. F. Sassen, "Wijsgerig leven in Nederland in de eerste helft der negentiende eeuw", *Mededelingen der Koninklijke Nederlandsche Akademie van Wetenschappen. Afdeling letterkunde.* Nieuwe Reeks 17 (1954), no. 10.

2. G.J.D.C. Stempel, "De eerste Kant-waardeering in Nederland", *De Gids* 86 (1922) II, 483-501; H.J. Groenewegen, "Kant in Nederland", in: *Kant's betekenis voor de wijsbegeerte* (Amsterdam, 1942) 5-20.

wijsbegeerte, en de geschiedenis van dezelve (Magazine for the critical philosophy, and its history; 1799-1803). The magazine was intended to make "the Dutch public ... more familiar with the Critique of the pure, and the practical, Reason, devised by the widely famous Thinker of our century".[3]

It is of interest that various Dutch scientists were adherents of Kant's philosophy.[4] Already in 1776 Deiman, known as an early Lavoisierian, was an adherent of Kant. Later he emphasized "the basic forces according to the Principles of Immanuel Kant".[5] In particular Deiman referred to the *Metaphysische Anfangsgründe der Naturwissenschaft* (1786). He accepted Kant's dynamical theory of matter, but restricted himself to merely repeating the philosophical concepts of the Königsberg philosopher, keeping them strictly separated from his own important chemical work. Later Deiman published a book entitled: *De Geest en Strekking der critische Wijsbegeerte, in een kort overzicht samengesteld* (The Spirit and Tenor of the Critical Philosophy, composed in a short outline; 1805).

Deiman was not the only Kantian in the Dutch scientific community. Lambertus Bicker, physician in Rotterdam and secretary of the "Bataafsch Genootschap der proefondervindelijke Wijsbegeerte" (Batavian Society for experimental Philosophy) followed Deiman immediately after reading a memoir of his Amsterdam colleague "on the probability of a prime matter, of which light, heat, electricity, and magnetism are only special modifications" (1788).[6] The Amsterdam mathematician and philosopher Johan Frederik Lodewijk Schröder was also a Kantian[7] although he later abandoned Kant's ideas because of his deeply rooted

3. *Magazijn voor de critische wijsbegeerte* I (1799) II.

4. See my: "De ontvangst van Kant bij enige Nederlandse natuurwetenschapsbeoefenaars", *Scientiarum Historia* 12 (1970) 23-38.

5. J.R. Deiman, "De grondkrachten, volgens de Beginzelen van Immanuel Kant", *Magazijn voor de critische wijsbegeerte* 3 (1800) I, 1-38. German translation in: *Archiv für die Physiologie* 6 (1805) 491-517. See my: "Jan Rodolph Deiman (1743-1808)", *Wissenschaftliche Beiträge der Martin-Luther-Universität Halle-Wittenberg* 1982/6 (E43) 154-164.

6. J.R. Deiman, "Over de waarschynlykheid eener algemeene hoofdstoffe, van welke het licht, het vuur, de electriciteit en de magneetstoffe slegts byzondere wyzigingen zyn", *Algemeen Magazyn van Wetenschap, Konst en Smaak* 3 (1788) II, 931-954. See my: "Lambertus Bicker (1732-1801), An early adherent of Lavoisier in The Netherlands", *Janus* 67 (1980) 101-123.

7. J.F.L. Schröder, "Over de ruimte", *Magazijn voor de critische wijsbegeerte* 2 (1799) 275-304; "Proeve, om de werking der uitbreidende en aantrekkende kracht, door de zwaartekracht, en de form der lighaamen, aantetoonen", *Ibid.* 3 (1800) II, 102-127, and 4 (1801) 108-140.

Christian faith. Around 1800 he rejected the atomistic and accepted the dynamical system, applying this to explain the impenetrability of water, its different kinds and densities, etc. Also physicians like Jacob Elisa Doornik of Amsterdam and Willem Servaas of Delfshaven advocated Kant's philosophical ideas. Analysis of writings of Dutch Kantian scientists show that, although he had his adherents in the Netherlands at an early stage, Kant exercised but little influence on the practical study of the natural sciences. His general criticism was adhered by natural scientists and his dynamical theory of matter was accepted, without, however, any application to the natural phenomena. His adherents kept themselves to reflecting Kant's ideas; they did not apply them to explain scientific facts. In 1809 the Kantian philosopher Jacob Nieuwenhuis noticed that in Germany after Kant purely transcendental systems were built. He considered himself fortunate

> "to belong to a nation, which also in arts and sciences proceeds along the way of solid erudition with a quiet and certain step, and which does not, like the inhabitants of more southern regions, have the habit of flying on the wings of fiery imagination over all kinds of precipices and dangerous abysses."[8]

In 1807 Nieuwenhuis was, like Kant, of opinion that the principles to serve as fundaments for a scientific theory can not be deduced from experience, but a priori from the laws of our intellectual capacity.[9] Nieuwenhuis rejected both atomism and Kant's dynamism. He seemed to have a greater appreciation for the "genetic, constructing, or organic system of Schelling".[10] He wanted to construct Nature as a whole on Schelling's three categories of matter, namely magnetism, electricity, and galvanism. But in 1809 Nieuwenhuis's sympathy for Schelling had completely disappeared in favour of Kant's critical considerations.

While a number of Dutch scientists were early adherents of Kant's philosophical ideas, the influence of the *Naturphilosophical* systems of Friedrich Wilhelm Joseph von Schelling and Georg Wilhelm Friedrich Hegel was less noticeable in the Netherlands.[11] Obviously the sober minded Dutch scientists could not easily accept speculative systems which the German chemist Julius Liebig called in 1840 "an Worten und Ideen so reiche, am wahren Wissen und gediegenen Studien so arme".[12]

8. J. Nieuwenhuis, *Verhandelingen uitgegeven door Teyler's Tweede Genootschap* 16 (1809) 81-256, on p. 238.

9. J. Nieuwenhuis, *Schouwburg van in- en uitlandsche letter- en huishoudkunde. Mengelwerk* (1807) 1-12; 264-285; on p. 7.

10. *Ibid.*, 10.

11. See my: "Romantiek en natuurwetenschappers in Nederland (1800-1850)", *De Negentiende Eeuw* 8 (1984) 72-90.

12. J. Liebig, *Reden und Abhandlungen* (Leipzig, Heidelberg, 1874) 34.

In 1816 Doornik stated that he was still an adherent of dynamism, now adding that he had chosen "the points of view of the so-called Naturphilosophie".[13] An extensive "general survey of the most prominent and most important tenets of natural philosophy", based on Christoph Wilhelm Hufeland's writings, was published in 1810 by the Amsterdam physician Frans van der Breggen Corneliszoon.[14] With this publication van der Breggen made German Romantic *Naturphilosophie* generally known within the Dutch community of physicians. It is remarkable that the relatively small number of adherents of Romantic *Naturphilosophie* were physicians. An interesting example is Conrad Gerard Ontijd, physician at The Hague. In 1801 he published two articles, both entitled "on the influence of chemistry on the actions of the animal body".[15] Apparently Ontijd did accept the *Naturphilosophical* antagonism between attracting and repulsing basic forces in nature, but he refused to use a vital force which is no more than a modification of two basic forces. In living nature a vital force is acting as a particular force. It is important to note that although Ontijd adhered to *Naturphilosophie* in medical practice he left the final word to empiricism. He attached great value to the application of chemistry to medicine, but he pointed out that experimental chemistry does not explain the principle of life.

As in the reception of Kant, also in the reception of Romantic *Naturphilosophie* the ideas of the German *Naturphilosophen* were merely repeated without any further elaboration. In fact the discussion was mainly confined to the problem of vital force, as introduced by Johann Christian Reill. The Delft physician Abraham van Stipriaan Luiscius, who also biassed in favour of Schelling's *Naturphilosophie* (1803), assumed, like Ontijd, an independent principle of life. Doornik on the contrary, defended the thesis that there is no particular principle of life.[16] The Kantian Doornik reduced everything in organic as well as in inorganic nature, to the two basic Kantian forces. Ontijd and van

13. J.E. Doornik, "Over het zoogenaamd dierlijk magnetismus", in: *Natuurwijsgeerige verhandelingen* (Amsterdam, 1816) 11.

14. F. van der Breggen, "Algemeen overzigt der voornaamste en meest-belangrijke grondstellingen der natuurwijsbegeerte", *Bijdragen tot de theoretische en praktische geneeskunde* 1 (1810) II, 96-124.

15. C.G. Ontijd, "Proeve over den invloed der Scheikunde op de werkingen des Dierelijken Lighaams", *Geneeskundig Magazijn* 1 (1801) I, 1-58 and III, 1-102. See my: "De Haagse arts Conrad Gerard Ontijd (1776-1844) en de 'Naturphilosophie'", in: A.H.M. Kerkhoff, A.M. Luyendijk-Elshout and M.J.D. Poulissen eds., *De Novis Inventis, Essays in the History of Medicine in honour of Daniel de Moulin* (Amsterdam, Maarssen, 1984) 413-436.

16. J.E. Doornik, *Verhandeling over de levens-kragt, volgens dynamische grondbeginzelen* (Amsterdam, 1802) 98-121.

Stipriaan Luiscius stressed the need of a particular and absolute vital force in explaining the phenomena of living nature.

Although the philosophical system of Kant and the *Naturphilosophen* was accepted by a number of Dutch scientists and physicians, it exercised no influence on the natural sciences themselves. When the young Leiden student Hendrik Carel van der Boon Mesch used *Naturphilosophical* ideas in an essay: *Commentatio chemica de ratione qua servat Natura proportionem inter principia proxima atmosphaerae* (1817), he was immediately seriously warned by Jacob Gijsbertus van Breda, then professor of botany, chemistry and pharmacy at the Franeker Athenaeum. "We warn him only for a word used in this meaning, borrowed from the jargon of German *Naturphilosophie*".[17]

The significance of Kantian and Romantic *Naturphilosophical* science and medicine in the Netherlands was marginal. In fact Dutch scientists restricted themselves mainly to accepting the idea of the unity of nature and to the use and interpretation of vital force. No doubt the reason has to be sought in the sober mind of the Dutch scientist at that time. He was interested in the practical application and elaboration of scientific theories based on experiments and not on speculations as in *Naturphilosopie*. Characteristic in this respect is a verdict on the German idealistic systems of his time by the experimental physicist Pieter Otto Coenraad Vorsselman de Heer:

> "In physics we acknowledge no other method of teaching than empiricism, and he who seeks a different one, is a false teacher and lost are those who hear him ... You shall know the tree by its fruit; it is the old blasted tree, planted by Aristotle, which has never brought forth anything."[18]

Professor Gregory's correct statement "about the impact of Kant's critiques on the German intellectual community at the end of the eighteenth and beginning of the nineteenth centuries" also applies to the Netherlands. His answer to the question "why Schelling was preferred by many scientists in the early years of the century" is correct as far as the preference of many German scientists is concerned, but does not stand for their Dutch colleagues. While Kant's philosophy of nature influenced a great number of Dutch scientists, this was not the case with Schelling's *Naturphilosophie*. In the Netherlands proper *Naturphilosophen* were white crows, no doubt because few scientists cared for the purely *Natur-*

17. J.G.S. van Breda, *Aanmerkingen omtrent den staat en de evenredigheid der bestanddeelen van de dampkringslucht, en onderzoek van hetgeen daaromtrent gesteld is in zekere Verhandeling van H.C. van der Boon-Mesch* (Amsterdam, 1820) 30.

18. B.W.A.E. Sloet tot Oldhuis, *Levensschets van P.O.C. Vorsselman de Heer, hoogleraar in de wis- en natuurkunde te Deventer* (Deventer, 1843) 109-110.

philosphical way of speculative natural science.

Do we here have an example of a natural style of scientific study? Whether or not we can speak of a distinctly Dutch style of science has hardly been investigated. Until now we can only establish the interesting and remarkable fact that Kant's ideas were acceptable to a number of Dutch scientists, but that the more speculative ideas of Schelling and others were evidently not. We still have the fascinating task to investigate the typical character, assuming there is one, of the Dutch way of studying natural phenomena.

University of Utrecht

Karl Hufbauer

SOLAR PHYSICS' EVOLUTION INTO A SUBDISCIPLINE (1945-1975)*

In 1967 a new scientific periodical — *Solar Physics: A Journal for Solar Research and the Study of Solar Terrestrial Physics* — made its debut. The journal, which was edited by C. de Jager of Utrecht and Z. Svestka of Ondrejov, Czechoslovakia, and published by D. Reidel Publishing Co. of Dordrecht, was a great success. Within five years, the press run climbed to 1,500 copies and the annual output to six volumes. The rapid rise of this journal was not an isolated phenomenon within the field of solar physics. At this very time, other subdisciplinary institutions — most notably, the European Joint Organization for Solar Observations (1968-) and the American Astronomical Society's Solar Physics Division (1970-) — were also making their appearance. Indeed, these developments within solar physics were but part of a much broader process of subdisciplinary formation that occurred throughout the republic of learning during the 1960s and 1970s.

As a means of illuminating solar physics' evolution into an astronomical subdiscipline and the broader concurrent process of subdisciplinary formation, I shall examine the background and success of the journal *Solar Physics*. In particular, I shall consider four questions. (1) How, during the three decades following World War II, did the international solar physics community attain the size needed to maintain a thriving periodical? (2) Why did de Jager undertake the task of organizing *Solar Physics*? (3) How did he and his co-editor Svestka go about making the journal into the community's forum between 1967 and 1975? And, by way of conclusion, (4) how have lean budgets since 1975 kept the solar physics community from becoming overly isolated within its new subdisciplinary institutions?

*Acknowledgments: This paper is an offshoot from a book on the history of solar physics that I am writing with the sponsorship of NASA's History Office. C. de Jager, L. Goldberg, M. van Hoorn, H.A.M. Snelders, and Z. Svestka kindly provided me with information and suggestions.

Boom Times in Solar Physics (1945-1975)

During World War II, statesmen, officers, bureaucrats, and managers all around the world became keenly aware of the potential of science for contributing to national might, prosperity, and health. In the ensuing decade, confronted by the problems of economic recovery and superpower rivalry, they assigned science a higher priority than ever before in their budgeting. But the postwar burgeoning of science was not due solely to new funds. Two other factors were also crucial. First, many scientists came away from the war with an opportunistic mind-set that enabled them to make the most of their new situation. In particular, they were eager to exploit wartime skills and contacts, willing to forsake prewar endeavors for more promising lines of inquiry, and readier than before to enter into team projects. And second, thanks to the war's stimulation of science-based technology, especially electronics, scientists were able to assemble or purchase better instruments for pursuing their research objectives.[1]

Solar physicists participated in science's postwar boom. Their first priority was to revive cooperative programs for monitoring the sun that had been established in the interwar era. They had much assistance in this endeavor from national agencies responsible for predicting the influence of solar phenomena on radio communications. By 1947, observatories around the globe were again sending solar statistics to Zürich for publication in the International Astronomical Union's *Quarterly Bulletin on Solar Activity*. Meanwhile, motivated not only by their concern with monitoring but also by their interest in investigating one or another specific phenomenon, solar physicists were improving their observing capabilities. Initially they could do little more than refurbish or upgrade instruments that had survived the war. Through 1948, for instance, they had to get by with pre-1945 coronagraphs. Thereafter, however, they found it easier to obtain funds for new spectrographic equipment, coronagraphs, and monochromatic filters.[2]

The outstanding participants in the postwar drive to expand solar

1. For the effects of World War II on science, see, for instance, D. Kevles, *The Physicists* (New York, 1979); A.C.B. Lovell, "The Effects of Defence Science on the Advance of Astronomy", *Journal for the History of Astronomy* 8 (1977) 151-173; and D.H. DeVorkin, "Electronics in Astronomy: Early Applications of the Photoelectric Cell and Photomultiplier for Studies of Point-Source Celestial Phenomena", *Proceedings of the IEEE* 73 (1985) 1205-1220.

2. For a survey of the equipment at the world's fifty observatories and stations conducting visual studies of the sun in 1953, see R. Coutrez, "Table of Solar Optical Instruments", in: G.P. Kuiper ed., *The Sun* (Chicago, 1953) 728-733.

observing capabilities were in their thirties at war's end. Not yet fully established, such men tended to be more alert to emerging governmental sources of patronage than were the leaders of the dominant prewar solar observatories. Moreover, they often had a greater awareness of technical opportunities for improving instrumentation. By 1957 several of these entrepreneurs — L. Goldberg (1913-1987), Chair of the University of Michigan's Department of Astronomy; W.O. Roberts (1915-), Director of Colorado's High Altitude Observatory; J.W. Evans (1909-), Superintendent of the Air Force's Sacramento Peak Observatory in New Mexico; A.B. Severny (1913-1986), Director of the Soviet Academy's Crimean Astrophysical Observatory; and K.O. Kiepenheuer (1910-1975), Director of the University of Freiburg's Fraunhofer Institute — had given their observatories important places in the gazetteer of solar physics.[3]

While solar physicists were resuming solar monitoring programs and developing more sophisticated versions of prewar instruments, outsiders were significantly extending the kinds of solar phenomena that could be observed. In particular, physicists and engineers were drawing on the technological and organizational legacies of World War II to open up new regions of the solar spectrum. Some — mostly men with experience in radio and radar projects — followed up on the wartime discovery that the sun emits radio waves with studies of its radio spectrum and sources of radio activity. They soon discovered that the sun is not the only discrete radio source in the heavens. Nonetheless, even if they were primarily interested in the non-solar sources, they found that they could often develop valuable techniques in or raise needed funds from their solar work. By 1953, therefore, groups at nineteen institutions were making at least intermittent radio observations of the sun. In fact, groups at ten of these institutions had sufficiently close contact with solar physics that they were sending regular reports to the *Quarterly Bulletin on Solar Activity*.[4]

3. All of these entrepreneurs except Roberts were asked to serve on the editorial board of *Solar Physics*. Although Roberts was not asked to join the board, two of his colleagues, R.G. Athay and J.W. Firor, Jr., were. See Appendix 2.

4. J.P. Wild, "Solar Radio Instruments", in: Kuiper ed., *The Sun*, 734-737. The group led by J.L. Pawsey in Sydney, Australia, was the pacesetter in solar radio astronomy through the 1950s. For the history of work there, see A.C.B. Lovell, "Joseph Lade Pawsey, 1908-1962", *Biographical Memoirs of Fellows of the Royal Society* 10 (1964) 228-243; and W.N. Christiansen, "The First Decade of Solar Radio Astronomy in Australia", in: W.T. Sullivan III ed., *The Early Years of Radio Astronomy: Reflections Fifty Years after Jansky's Discovery* (Cambridge, Eng., 1984) 112-131.

As some physicists and engineers were exploring the sun's radiation at radio wavelengths, others — primarily in the United States — were using rockets to gain access to the short end of the solar spectrum. In October 1946, a Naval Research Laboratory group led by R. Tousey obtained the first ultraviolet solar spectrogram with instruments carried aloft by a captured V-2. Thereafter, his and other teams working in military laboratories or with long-range military contracts developed a diverse array of instruments for securing ultraviolet and x-ray spectra from the sun, servomechanisms for pointing control, and electronic methods of data retrieval. Despite great technical virtuosity, the rocket results remained more promising than important through the 1950s because the spectral and spatial resolution was quite modest compared to that from ground-based telescopes. Still, the early rocket scientists created the basic instruments, techniques, and routines of what would soon become space solar physics.[5]

In the mid-1950s, solar physicists around the globe joined with geophysicists in preparing for the International Geophysical Year (IGY) which was planned to coincide with the maximum in solar activity during 1957-1958.[6] When observing commenced on 1 July 1957, ninety-five observatories and stations were set to participate in an ambitious program of solar monitoring. Twenty-six had the instruments and staff for monitoring three or more of the solar phenomena designated for coverage:

Observatories monitoring three or more IGY Solar Phenomena as of July 1957 (proceeding westward from Greenwich)

Great Britain: Greenwich-Herstmonceux (4), Edinburgh+ (4)
United States: Michigan-Pontiac+ (5), HAO Boulder*+ (3), HAO Climax (4), Sacramento Peak + (6), Mt. Wilson*+ (4), Honolulu-Makapuu Pt.+ (3)
Japan: Tokyo-Mitaka+ (4)
Australia: Sydney*+ (4)

5. For rocket solar physics, see R. Tousey, "Solar Work at High Altitudes from Rockets", in: Kuiper ed., *The Sun*, 658-676; and H. Friedman, "The Sun's Ionizing Radiations", in: J.A. Ratcliffe ed., *Physics of the Upper Atmosphere* (New York and London, 1960) 133-218. D.H. DeVorkin is writing an archivally-based history of rocket solar physics.

6. For the origins and planning of IGY, see H. Spencer Jones, "The Inception and Development of the International Geophysical Year", *Annals of the International Geophysical Year* 1 (1959) 383-413; and M. Nicolet ed., "The International Geophysical Year Meetings", *Ibid.* 2A (1958) 1-395.

India: Kodaidakanal (5)
Soviet Union: Tashkent (3), Kislovodsk+ (3), Moscow NIZMIR* (3), Crimea-Simeis*+ (4), Kiev (4)
Turkey: Istanbul (3)
Greece: Athens (3)
Czechoslovakia: Prague-Ondrejov+ (5)
Austria: Kanzelhoehe (5)
Federal Republic of Germany: Wendelstein (3), Freiburg*+ (3)
Italy: Florence-Arcetri*+ (4)
Switzerland: Zürich*+ (3)
Belgium: Uccle (4)
France: Meudon*+ (5)
* = World Data Center for one or more solar subjects;
+ = one or more staff members belonged to *Solar Physics*' editorial board during 1966-1968 (see Appendix 2).

(Source: "Catalogue of [Solar] Data in the World Data Centers", *Annals of the International Geophysical Year* 3 (1963) 243-269.)

Besides the optical and radio astronomers, the rocket scientists interested in the sun were also ready to undertake work. In fact, the Naval Research Laboratory group led by H. Friedman marked the start of observations by firing the first of fifteen Nike-Deacon rockets dedicated to the study of solar flares.[7]

Solar physicists hoped that the data collected during IGY would increase understanding of solar activity and its influence on terrestrial phenomena. The most immediate contribution came as a result of Friedman's rocket research on flares which confirmed the hypothesis that solar x-ray bursts are the cause of fadeouts in shortwave radio reception. As it happened, however, IGY's most important contribution to solar physics was an indirect, and in most circles completely unanticipated, one — Sputnik. Launched on 4 October 1957, the satellite was billed by the Soviet Union as a contribution to IGY. It was that, and much more. Sputnik was a hitchhiker on the third successful test of the new Soviet intercontinental ballistic missile. It was a declaration of the Soviet Union's plans to share in world leadership. It was a testament to modern dreams of space exploration. And it was a goad to the United States. Together with subsequent Soviet space firsts, Sputnik provoked a rapid buildup of American missile forces, motivated the creation of the National Aeronautics and Space Administration, and stimulated a major

7. L.V. Berkner ed., "Manual on Rockets and Satellites", *Ibid.* 6 (1958) 1-508, p. 70, 109.

expansion in funding for science and science-based technology. Nothing less, it seemed to American elites, would maintain the nation's security and prestige.[8]

By stimulating scientific competition between the superpowers and strenuous efforts by smaller nations to maintain their scientific standing, Sputnik engendered much new patronage for solar research. The scientists who translated this patronage into solar observing capabilities pursued three main strategies. Many, working at ground-based observatories, obtained new optical and radio telescopes or substantially upgraded existing instruments. Among their most notable achievements were the completion of the huge McMath Solar Telescope at the Kitt Peak National Observatory in 1962 and the first large vacuum tower telescope at the Air Force's Sacramento Peak Observatory in 1969. Others utilized stratospheric balloons, rockets, and especially spacecraft to develop space solar physics. Working mainly with NASA, the Naval Research Laboratory, and the Soviet Rocket Forces, they got ever more sophisticated payloads for observing the sun above the atmosphere. Their most ambitious mission was NASA's Skylab. Between May 1973 and February 1974, three successive crews used Skylab's eight solar telescopes to a store of data that was so immense and rich that it is still undergoing analysis. Finally, a very few researchers — all nuclear scientists — went underground, drawing on the new patronage to establish solar neutrino detectors in deep subterranean facilities.[9]

As solar observing evolved into an ever larger and more complex enterprise between 1957 and 1975, the number of solar physicists grew apace. Centrifugal tendencies within the field also became stronger. The triennial meetings of the International Astronomical Union's solar committees no longer sufficed to maintain the field's cohesion. Solar physicists needed subdisciplinary institutions both to nurture their mutual interaction and to represent their interests within the discipline as a whole. The journal *Solar Physics* was the first such institution to make its appearance. We turn now from its general context to its local background.

8. W.A. McDougall, *The Heavens and the Earth: A Political History of the Space Age* (New York, 1985); and H.E. Newell, *Beyond the Atmosphere: Early Years in Space Science* (Washington, DC, 1980) = NASA SP-4211.

9. For a general review of the resources available to solar physicists by 1975, see Federal Council for Science and Technology: Solar Astronomy Task Force, "Report to the Interagency Coordinating Committee on Astronomy" (June 1975). For Skylab's solar work, see J.A. Eddy, *A New Sun: The Solar Results form Skylab* (Washington, DC, 1979) = NASA SP-402; and W.D. Compton and C.D. Benson, *Living and Working in Space: A History of Skylab* (Washington, DC, 1983) = NASA SP-4208.

No one could have predicted at the end of World War II that Utrecht would emerge as a center for solar physics in the next two decades. To be sure, the physicist W.H. Julius (1860-1925) had inaugurated a tradition of solar research here at the beginning of the century.[10] To be sure as well, this tradition had been carried on with distinction by M.G.J. Minnaert (1893-1970), a Flemish emigré who worked in Utrecht's physics laboratory until his appointment as Director of the Sonnenborgh Observatory in 1937.[11]

Nonetheless, the prospects in 1945 for a postwar flowering of solar physics in Utrecht were distinctly unfavorable. As a left-leaning foreigner, Minnaert was an outsider who, despite more than a decade at the forefront of solar spectrophotometry, was still an extraordinary professor at age fifty-two. Moreover, the outlook for funding was bleak, partly because spending for science would necessarily remain depressed until the Netherlands recovered from the ravages of war and partly because Leiden, the traditional leader of Dutch astronomy, was likely to be given first crack at scarce resources. Finally, Utrecht was not favored by nature with exceptional conditions for observing the sun.

Undaunted by the poor prospects, Minnaert set about trying to develop solar physics at Utrecht after the war. His odds of success were improved in 1946-1947 when Utrecht University advanced him to full professor and the Royal Astronomical Society awarded him its coveted gold medal. Thereafter he gradually enlarged the observatory's faculty and scientific staff and recruited a talented coterie of advanced students (see Appendix 1). He directed most of the research toward the interpretation of solar spectra that had been acquired from better-located observatories or during eclipse expeditions. Minnaert's success in creating a lively group that was at the vanguard in solar spectroscopy was manifest at an informal meeting held in Utrecht in mid-1956. One of those present — Harvard's R.N. Thomas — reported that the "Utrecht meetings were shorter [than Liege's Seventh International Colloquium on

10. M.G.J. Minnaert, "Julius, Willem Henri", in: C.C. Gillispie et al. eds., *Dictionary of Scientific Biography* (New York, 1973) VII, 186-187; H.A.M. Snelders, "Julius, Willem Henri", in: J. Charité et al., eds. *Biografisch Woordenboek van Nederland* (Amsterdam, 1985) II, 270.

11. C. de Jager, "Herdenking van Marcel Gilles Jozef Minnaert", Koninklijke Nederlandse Akademie van Wetenschappen: *Jaarboek* (1970-1971) 234-239, and "Marcel Gilles Jozef Minnaert (1893-1970)", American Philosophical Society: *Year Book* (1971) 134-138; L. Buning, "Minnaert, Marcel Gilles Jozef" in: J. Charité et al., eds., *Biografisch Woordenboek van Nederland* ('s-Gravenhage, 1979)I, 391-393.

molecules in stars], but of course more packed, minute for minute, with things of immediate and pressing interest. No abstracts, no records, much opinion, much fun."[12]

Among the scientists at the Utrecht meeting, C. de Jager (1921-) — the subsequent organizer of the journal *Solar Physics* — made the biggest impression on Thomas. De Jager had matriculated at Utrecht University on the eve of World War II. He spent the war years here, partly in concealment, studying mathematics, physics, and his favorite subject, astronomy. In 1946, after brief service as an assistant in both theoretical physics at Utrecht and astronomy at Leiden, he accepted an assistantship in the Sonnenborgh Observatory. Two years later he launched his career as a solar physicist with a paper (on the solar atmosphere) in the Royal Dutch Academy's *Proceedings*. Four years later yet, he completed his dissertation on "the hydrogen spectrum of the sun." De Jager, already a productive author, accelerated his pace through the remainder of the 1950s, publishing twenty-one articles on solar physics, fifteen on stellar physics, and five on meteors and meteorites, a popular book on stars and planets, and, as his *pièce de resistance*, an immense treatise on "the structure and dynamics of the solar atmosphere" in the *Handbuch der Physik* (1959). Given this fertility and versatility, it is small wonder that de Jager impressed Thomas at the 1956 meeting. To judge from his record of advancement, he also impressed his mentor Minnaert. In 1951 he was promoted to scientific employee, in 1957 to lector (assistant professor) in stellar astrophysics, and in 1960 to professor of general astrophysics.[13]

Upon becoming Utrecht's second full professor of astronomy, de Jager took a close look at the options for the observatory's long-range development. He concluded, as a result of both his prior work on solar spectra obtained by American rocket scientists and his current membership on the Royal Dutch Academy's Committee for Geophysics and Space Research (established 5 March 1960), that the best strategy for Utrecht was to develop the capability of participating in the American and European space programs. In pursuit of this strategy, he first persuaded the Academy's committee of the importance of developing solar x-ray spectrographs for rockets and satellites and volunteered to head up the project. Then in 1961, with governmental funding arranged

12. Thomas to J.W. Evans and W.O. Roberts (24 July 1956), Roberts Papers, Library, University of Colorado, Boulder, CO, Box 14:2.

13. De Jager, Curriculum Vitae (1984), interview with K. Hufbauer (24 May 1985), and "Wereld en Wetenschap" (Afscheidsrede Rijksuniversiteit Utrecht, 24 April 1986).

by the committee, he inaugurated a process of subdivision within the observatory by founding the "division of space research" which evolved over the next four years into the "division for solar and stellar space research" and acquired its own laboratory.[14] Meanwhile in 1963, having proven himself as both a scientist and institution builder, de Jager succeeded Minnaert as the observatory's manager. During August, the succession was celebrated with a well-attended Symposium on the Solar Spectrum, organized by de Jager and dedicated to Minnaert.[15]

In 1960-1961, just when de Jager was concluding that the future of Utrecht astronomy depended on going into space research, A. Reidel was seeking ways to consolidate the recent transformation of his family's sixty-year-old printing firm in Dordrecht into a full-fledged publishing company. Reidel had acquired *Synthese: An International Quarterly for the Logical and Psychological Study of the Foundations of Science* from Kroonder in 1958 and was planning to start *Studies in Soviet Thought* in 1961. He recognized that the post-Sputnik space boom provided an opportunity for adding a new journal to his company's list. But an editor had to be found. After several refusals, Reidel asked de Jager whether he might consider the job. De Jager, thinking that this could be an excellent way to acquaint himself with the trends in and leaders of space research, indicated that he was interested. However, he had two conditions. He wanted to limit the journal to invited review papers. And he wanted the assistance of a distinguished editorial board. Reidel readily agreed. Thus was born *Space Science Reviews*, the first issue of which appeared in June 1962.[16]

Some three years later, impressed that this periodical with its run of 2,000 copies was twice the size of the other two journals on the firm's list[17], Reidel came to de Jager with a new proposal. The firm wondered if he could be persuaded to edit a second periodical on some branch of astronomy. De Jager thought solar physics would be the best choice for such a subdisciplinary journal. But he was reluctant to undertake the job. Besides his editorship of *Space Science Reviews*, he was orchestrating a rapid expansion of astronomy at Utrecht (see Appendix 1), serving as an

14. J. Veldkamp, Annual reports on the activities of the Nederlandse Commissie voor Geofysica en Ruimte-Onderzoek, Koninklijke Nederlandse Akademie van Wetenschappen: *Jaarboek* (1960-1961) 201-202, (1961-1962) 222-224, (1962-1963) 283-286, (1963-1964) 280-301, (1964-1965) 256-265; *Studiegids der Rijksuniversiteit Utrecht* (1962-63) 80, (1965-1966) 85; de Jager, "Wereld en Wetenschap".

15. De Jager ed., *The Solar Spectrum* (Dordrecht, 1965).

16. D. Reidel Publishing Company, *Guide for Authors and Editors* (Dordrecht, 1983) vi; De Jager, interview, and "Introduction", *Space Science Reviews* 1 (1962) 5-6.

17. *Handboek van de Nederlandse Pers* (March 1965) 19, 33, 36.

extraordinary professor in Brussels, participating in a growing number of international committees and commissions, and, remarkably, remaining active in research. Reidel persisted. On thinking the matter over, de Jager could see that by editing such a journal he would simultaneously buttress Utrecht's position within solar physics and help the field consolidate its position within astronomy. He yielded, therefore, to Reidel's entreaties.[18]

De Jager, Svestka, and the Rise of Solar Physics (1966-1975)

De Jager's first step was to canvass fifteen colleagues around the globe about the journal's prospects and organization. Most of his respondents were enthusiastic. Indeed, the new President of the International Astronomical Union's Commission 10 for solar activity — Z. Svestka (1925 -), a Czech specialist on solar flares[19] — reported that he had begun promoting just such a journal at the Union's 1964 congress in Hamburg. Despite their general enthusiasm, de Jager's respondents did anticipate three possible problems. First, it was likely that some important solar physicists would not contribute to the journal, lessening its value as a forum for the field. Second, there was a risk that some second-raters would bombard the journal with manuscripts, lowering its quality. The third potential problem was of a higher order. Papers of broad significance to astrophysics, especially theoretical studies of the solar atmosphere, would lose their impact because they would no longer be seen by stellar physicists. In March 1966, de Jager gave his reactions to these concerns in a circular letter inviting thirty-seven colleagues to join *Solar Physics*' editorial board. He thought that the journal's comprehensiveness could be assured by publishing abstracts of important work that appeared elsewhere. He was confident that high standards could be maintained if editorial board members would be strict in their refereeing. And, while conceding that theoretical work of broad applicability "should better be published in general astrophysical journals," he believed that the journal would be making a real contribution if it brought together work that focused on the sun, particularly "solar activity in all its forms."[20]

18. De Jager, Curriculum Vitae, and "Wereld en Wetenschap"; *Studiegids der Rijksuniversiteit Utrecht* (1965-66) 85; de Jager and Z. Svestka, "Editorial", *Solar Physics* 1 (1967) 3-4.

19. Svestka, Curriculum Vitae (1984).

20. De Jager to potential editorial board members (16 March 1966), L. Goldberg Papers, Tucson, AZ.

De Jager's colleagues responded warmly to his invitation. Within three months, all but three of those on his initial list agreed to serve on *Solar Physics*' editorial board. Moreover Svestka, who had been sent a special invitation to share in the journal's direction, welcomed the opportunity to help provide solar physicists from the socialist nations with a truly international forum for their work. In June 1966, de Jager and Svestka took advantage of a colloquium on solar fine structure at the Fraunhofer Institute's station on Capri to discuss plans for the journal with several board members. They also recruited two additional members. Shortly after his return to Utrecht, de Jager informed those who had agreed to serve on the editorial board of Svestka's acceptance of the co-editorship. In addition he reported that an announcement of the journal and a request for manuscripts would soon be sent out to the entire solar physics community. His hope was that the first issue would appear in about six months.[21]

De Jager and Svestka's letter to potential contributors was entirely upbeat. They made no concession whatsoever to the idea that *Solar Physics* was inappropriate for articles that might be of interest to stellar physicists. Instead, they promised that

> "SOLAR PHYSICS will deal with all aspects of solar problems, from the interior to the solar wind in interplanetary space. It will also deal with sun-earth relations [in those] cases where the emphasis is solar, for instance when the earth's atmosphere or magnetic field is used as a spectrum analyzer for the solar electromagnetic or particle spectrum. SOLAR PHYSICS will contain theoretical, observational and instrumental papers, and will refer to ground-level observations as well as to those made from balloons or space vehicles, by means of radiotelescopes, and riometers, neutron monitors and other ground-level indicators of solar particle activity."[22]

Two months later, in August 1966, Reidel began advertising for subscribers. The firm proclaimed that *Solar Physics* would

> "for the first time provide solar physicists with an exclusive medium for the presentation of their work, since such information has hitherto been scattered over many scientific journals. Such concentration of research papers will profit not only

21. Svestka to Hufbauer (2 June 1986); De Jager and Svestka, "Editorial", *Solar Physics* 1 (1967) 3-4; K.O. Kiepenheuer ed., *The Fine Structure of the Solar Atmosphere* (Wiesbaden, 1966) 7-8; De Jager to editorial board members (13 June 1966), L. Goldberg Papers, Tucson, AZ. The three who did not accept de Jager's initial invitation were R.O. Redman, H. Friedmann, and R.B. Dunn. The two recruits at Capri were A. Bruzek and R. Giovanelli. Before 13 June, de Jager had also recruited R. Tousey.

22. De Jager and Svestka to potential contributors (June 1966), L. Goldberg Papers, Tucson, AZ. The list of editorial board members in this letter indicates that V.A. Krat had joined the editorial board since 13 June 1966.

the specialist but also the nonspecialist interested in the recent progress in solar physics."[23]

Confident of success, the firm soon reported in the trade press that it would be publishing 1,000 copies of the journal.[24]

The first issue of *Solar Physics* appeared almost exactly on schedule in early 1967. The editorial board, including the chief editors, numbered forty-one solar physicists from twelve nations around the globe (see Appendix 2). Speaking for the board, de Jager and Svestka expounded on the journal's rationale in their preface. Contemporary scientists were barraged, they observed, with an "ever growing stream of publications ... scattered over many different journals." In a "well-defined" field like solar physics, such a scattering of research was unnecessary. The new journal would ease the work of solar physicists by providing them with a "vehicle where results ... could be found together." The ultimate "objective" was to "establish a closer cooperation between all those concerned with solar research." In short, the editors and their board believed that *Solar Physics*, by serving the booming field as a forum, would facilitate research and foster cooperation.[25]

To the delight of publishers and editors alike, *Solar Physics* flourished. Circulation grew to the point that Reidel was obliged to increase the press run to 1,200 copies in early 1969 and on to 1,500 copies in mid-1971.[26] Meanwhile, de Jager and Svestka were under increasing pressure from submissions. As early as the spring of 1967, they could see that eight issues a year would be needed to keep up with the influx of manuscripts. Half a year later, they realized that they could soon shift to monthly publication. Upon reporting this happy development to the editorial board, they also sought to increase its attachment to the journal by having an election for five new members.[27] Even monthly publication did not suffice. By November 1968, de Jager and Svestka had to announce a 34% increase in the number of pages per monthly issue. In July 1969, they coupled their next announcement of an increase in issue size with a report that referees had been asked to judge the quality and length of manuscripts with greater severity. The referees were tougher, boosting

23. For a sample advertisement, see *Space Science Reviews* 5 (Aug 1966).

24. *Handboek van de Nederlandse Pers* (Sep 1966) 22.

25. De Jager and Svestka, "Editorial", *Solar Physics* 1 (1967) 3-4.

26. *Handboek van de Nederlandse Pers* (Sep 1968), 23, (March 1969) 24, (March 1971) 27, (Sep 1971) 26.

27. De Jager and Svestka, "Editorial", *Solar Physics* 1 (1967) 303, and letter to editorial board members (3 Oct 1967), L. Goldberg Papers, Tucson, AZ. For the election's outcome, see Appendix 2.

rejections from the 6% rate of 1966-1967 to 11% by 1971. Nonetheless, de Jager and Svestka had to increase issue size yet again in late 1971 in order to reduce the swelling backlog of accepted articles. The consequence was that the annual number of volumes climbed to six and the cost of an individual subscription to just under $100 a year. In July 1972, eager to avoid any further increases, de Jager and Svestka again returned to the problems of growth and backlogs in an editorial. Their solution was to announce that henceforth short articles would be given precedence over those longer that ten printed pages.[28]

That fall ten solar physicists at Sacramento Peak Observatory, believing that *Solar Physics* had already become too large and expensive, urged de Jager and Svestka to reconsider the problem. They suggested several means of reducing costs to individual subscribers: introducing page charges; increasing the differential between institutional and individual subscriptions; raising refereeing standards to the level of the *Astrophysical Journal*; limiting free distribution to the most active referees; and printing on cheaper paper. Several, but not all, of the rebels argued that page charges would be the best remedy — such fees would not only fund lower subscription prices but also discourage verbosity. Evidently worried that their advocacy might be dismissed, this group informed de Jager and Svestka that the matter would be raised at the January 1973 meeting of the American Astronomical Society's Solar Physics Division.[29] De Jager was quick to respond. He indicated that he too was concerned about the journal's cost to individual subscribers. He and Svestka feared, however, that the introduction of page charges would turn *Solar Physics* into an American journal since American institutions were the only ones willing to pay such fees. In concluding, he promised to discuss the matter at length with Svestka, who was then at the Fraunhofer Institute in Freiburg, and Reidel.[30]

Although de Jager and Svestka obtained a substantial decrease in the price of individual subscriptions by February 1973, the concerns about *Solar Physics*' practices, once having been articulated, were not so easily laid to rest. In particular, the discussion of the journal at the Solar Physics Division's meeting seems to have engendered further criticism of

28. De Jager and Svestka, Editorials, *Solar Physics* 5 (1968) 254; 8 (1969) 3; 22 (1972) 3-4.

29. D.M. Rust, R.C. Altrock, R.B. Dunn, S. Musman, R. Canfield, and W.J. Wagner to de Jager (6 Nov 1972); and G.W. Simon to de Jager and Svestka (8 Nov 1972), Director's Papers (1973), High Altitude Observatory, Boulder, CO.

30. De Jager to Rust et al. with copies to the editorial board (27 Nov 1972), Director's Papers (1973), High Altitude Observatory, Boulder, CO.

its standards. Following up on the discussion, G.A. Newkirk, Jr., the division's outgoing chairman, an elected member of the editorial board, and the Director of Boulder's High Altitude Observatory, asked a stellar theorist for an evaluation of *Solar Physics*. He was advised that, though the journal carried many fine papers, its acceptance threshold was still considerably below that of the *Astrophysical Journal*. Newkirk, who was just then marshaling outside support for his observatory which was under external review, may have seen some advantage in championing high standards. In any case, he sent a circular letter to the editorial boards' members, reporting his concern that some solar as well as non-solar colleagues were of the opinion that "we include too many papers which are not worthy of publication in a refereed journal." He urged that those who shared his concern about the journal's future join him in requesting the editors to set up a board meeting at the International Astronomical Union's congress in Sydney that August.[31]

The board's response to Newkirk's letter was sufficient to induce de Jager and Svestka to call the meeting. They "were distressed" that several board members thought some bad papers were appearing in *Solar Physics* and were eager to develop better guidelines for referees. They also believed that steps should be taken to strengthen the board. Perhaps, they suggested, the board's members should serve staggered terms, retiring after four years unless they were re-elected by the board. The editors-in-chief, they also suggested, should serve staggered eight-year terms. De Jager, indeed, volunteered to step down at the end of 1974 and Svestka, four years later.[32]

At the editorial board's meeting in Sydney on 25 August 1973, de Jager first discussed the finances of *Solar Physics* with the seventeen members present. Then he reported the startling fact that Newkirk's circular letter had caused the rejection rate to jump suddenly from 12% to 30% and remain there. Those present agreed that a rate of 25% should be regarded as the journal's new norm. They also agreed that referees should be asked to judge papers on their originality, brevity, and physical coherence. And, after persuading de Jager to remain at the helm, they reached a consensus that board members should be elected for staggered terms.[33] The editorial

31. D. Mihalas to Newkirk and J.B. Zirker (22 Feb 1973); Newkirk to Roberts regarding the forthcoming review (10 April 1973); and Newkirk to editorial board members with copies to the executive committee of the Solar Physics Division, [12-13 April 1973], all in Director's Papers (1973), High Altitude Observatory, Boulder, CO.

32. De Jager and Svestka to editorial board members [ca. 25 May 1973], Director's Papers (1973), High Altitude Observatory, Boulder, CO.

33. Newkirk?, "Short report of the meeting of the members of the Editorial Board

board's first regular election was held during 1974. Over ninety percent of the board members joined de Jager and Svestka in balloting for the eleven positions that were due to fall vacant at year's end. The upshot was rejuvenating, while the board as a whole had a median age of 54 at the time of the election (see Appendix 2), the eleven new and re-elected members had a median age of 40.[34]

Between 1966 and 1975, therefore, de Jager and Svestka presided over the development of *Solar Physics* into a successful forum for the international solar physics community. One indicator of their success was the journal's large circulation. Another was the journal's publication of over two hundred articles a year on all aspects of solar physics. The most telling indicators, however, were the row over standards and the institution of regular elections for editorial board members. These episodes revealed that leading solar physicists had come, as de Jager must have hoped from the outset, to regard the journal's performance as a collective responsibility.

Tempering of Separatist Inclinations (1975-)

By the mid-1970s, solar physicists constituted a small, yet robust subdisciplinary community at the boundary between astronomy and physics. They had, as we have seen, a thriving international journal with an elected editorial board. They also had, as I shall discuss elsewhere, two to four major conferences a year and a medley of organizations to represent their interests in disciplinary societies and before funding agencies. The community's only lack, and it did not seem terribly serious in that era of abundance, was a solid base in academia.

While most solar physicists valued the easy access that their new subdisciplinary institutions gave them to peers who shared their fascination with the sun, a few worried that the field was becoming too

of SOLAR PHYSICS" (25 Aug 1973), Director's Papers (1973), High Altitude Observatory, Boulder, CO; and de Jager, "Report from Editorial Board Meeting", *Solar Physics* 34 (Feb 1974) 275.

34. De Jager and Svestka to editorial board members (14 Mar, 30 Sep, 4 Dec 1974), L. Goldberg Papers, Tucson, AZ. The rejuvenation occurred partly because several senior members of the board, Biermann, Brueck, Goldberg, Houtgast, and Oehman, decided to stand aside for younger workers and partly because most of the new nominees were under 40. The eleven elected were J.M. Beckers+ (1934-), J.W. Harvey (1940-), C. Jordan (1941-), J. Kleczek+ (1923-), M. Kuperus (1939-), W.C. Livingston (1927-), F. Meyer, E.H. Schroeter (1928-), J.O. Stenflo (1942-), W.E. Stepanov+, and P.A. Sturrock (1924-). + indicates the person was re-elected.

isolated from the mainstream of astronomy. One such was L. Goldberg, the Director of Kitt Peak National Observatory and a retiring member of *Solar Physics'* editorial board. In the fall of 1974, he organized a little study of citation practices by authors publishing in the *Astrophysical Journal.* He found that the 30 solar papers appearing in this journal between 1 May and 15 October 1974 cited papers in *Solar Physics* 148 times, but that all the remaining papers, about 450, only did so 19 times.[35] Although Goldberg seems not to have done anything with this evidence of solar physics' isolation, other worriers were more outspoken. In 1976, for instance, J.C. Pecker and R.N. Thomas, both of whom had attended Utrecht's informal solar meeting back in 1956 and now worked at the Institut d'Astrophysique in Paris, critiqued the growing insularity of solar physics and other astronomical subdisciplines:

> "one sees strange attempts to emphasize the *diverse* aspects of solar-stellar-galactic studies ... rather than the *common* aspects. One studies solar, *or* stellar, *or* galactic astrophysics: one "ghetto-izes" his narrow subject; and lives apart from, not with, the others ... the exclusive concentration upon narrow areas reflects the specialist's feelings that he must develop his subject in exhaustive detail: but he risks confusing over-computed model and under-physiked reality because of neglected anomaly."

Urging a "'Jericho' outlook", Pecker and Thomas went on to point out the many advantages of knocking down the walls between solar and stellar research. They did not refer specifically to *Solar Physics.* But there is no mistaking their belief that the journal had helped ghettoize the field.[36]

In the mid-1970s, solar physics moved from an era of abundance into one of comparative scarcity. Funds needed for space solar missions and the support of ground-based solar programs became increasingly difficult to obtain. In part, the scarcity was a result of the stringent scientific budgets accompanying the hyper-inflation that followed the Vietnam War and energy crisis. In part, it was a consequence of the field's poor competitive position vis-a-vis those astronomical subdisciplines with stronger bases in academia. One response of solar physicists to their straitened circumstances was a tempering of their separatist inclinations. More so than in the 1960s and early 1970s, they were alert to opportunities for demonstrating their field's value to the astronomy as a

35. Goldberg, "Astrophysical Journal Survey", manuscript [ca. Nov 1974], L. Goldberg Papers, Tucson, AZ.

36. Pecker and Thomas, "Solar Astrophysics: Ghettosis from, or Symbiosis with, Stellar and Galactic Astrophysics?", *Space Science Reviews* 19 (1976) 217-243, quotation 223. This article was written at the invitation of de Jager, who was also eager to stimulate greater interaction between solar and stellar physicists; see de Jager to Hufbauer (26 May 1986).

whole. Partly for this reason and partly as well because of labor slowdowns at Reidel Publishing Co., many solar physicists worked harder at placing their articles in disciplinary journals. Besides disseminating their work more widely, some solar physicists have joined with stellar physicists in exploring and developing solar-stellar connections. Thus, the stringencies since the mid-1970s have had the salutory effect of encouraging members of the solar physics community to look for audiences and problems beyond as well as within the confines of their new subdiscipline.

University of California, Irvine, and NASA

APPENDIX 1
FACULTY, STAFF, AND DOCTORATES AT THE UNIVERSITY OF UTRECHT'S OBSERVATORY/ASTRONOMICAL INSTITUTE, 1945-1975

1945-46: Prof. extr. M.G.J. Minnaert*, director; 1 assistant. Doctorate in last 5 years: 1942: J. Houtgast*.

1950-51: Prof. M.G.J. Minnaert*, director; 2 scientific staff. Doctorates in last 5 years: 1946: H.C. van de Hulst; 1947: G.C.J. Zwanikken.

1955-1956: Prof. M.G.J. Minnaert*, director; J. Houtgast*, lector; 2 scientific staff. Doctorates in last 5 years: 1951: W.J. Claas*; 1952: C. de Jager*; 1955: P.J. Gathier.

1960-61: Prof. M.G.J. Minnaert*, manager; C. de Jager*, professor; J. Houtgast*, lector; 6 scientific staff. Doctorates in last 5 years: 1959: J. van Diggelen; 1960: F. van 't Veer, H. Hubenet*.

1965-66: Prof. C. de Jager*, manager; H.G. van Bueren and A.B. Underhill, professors; J. Houtgast*, lector; 17 scientific staff; 3 non-university scientific staff. Doctorates in last 5 years: 1964: A. Schadee*, J.M. Beckers*; 1965: M. Kuperus*, J.R.W. Heintze*, C. Zwaan*.

1970-71: Prof. C. de Jager*, manager; H.G. van Bueren, professor; E.A.M. Mueller*, visiting professor; A.D. Fokker*, W. de Graaff, J. Houtgast*, and C. Zwaan*, lectors; 31 scientific staff; 2 nonuniversity scientific staff. Doctorates in last 5 years: 1966: T. de Groot*, L.D. de Feiter*, J. Haisma, H. de Lang; 1968: E.P.J. van den Heuvel, J. Roosen*; 1969: R. Mewe, M.J.H. de Groot, C.J.E. Schuurmans*; 1970: H. Nieuwenhuijzen, W. van Rensbergen.

1975-76: Prof. C. de Jager*, manager; H.G. van Bueren and M. Kuperus*, professors; L.D. de Feiter* (d. 1975), A.D. Fokker*, W. de Graaff, A.F. Hearn, J. Houtgast*, C. Zwaan*, lectors; 37 scientific staff; 5 non-university scientific staff. Doctorates in last 5 years: 1971: J. van Nieuwkoop*; 1972: A.C. Brinkman*, R.C.P. van Helden; 1973: H.F. van Beek*, J. Rosenberg*; 1974: H. van de Stadt, H.J.L.M. Lamers; 1975: M.W.M. de Grauw, J.M.E. Kuijpers*, P. Hoyng*.

* = research mainly in solar physics.

Summary

	1945	1950	1955	1960	1965	1970	197
Profs and Lectors	1	1	2	3	4	7	9
Scientific Staff	1	2	2	6	20	33	42
Doctorates in preceding 5 yrs	1	2	3	3	5	11	10

(Sources: Faculty and Staff: *Gids der Rijksuniversiteit Utrecht* (1946-1957); *Studiegids der Rijksuniversiteit Utrecht* (1958-1968); *Gids Utrechtse Universiteit* (1969-1975); Doctorates: M. van Hoorn to K. Hufbauer, 19 febr. 1986.)

APPENDIX 2
SOLAR PHYSICS' EDITORIAL BOARD, 1966-1968

(1) Alfven, H. (1908-), Stockholm, Sweden
(1) Athay, R.G. (1923-), Boulder, CO, USA
(1) Beckers, J.M. (1934-), Sacramento Peak, NM, USA
(1) Biermann, L. (1907-1986), Munich, German Federal Republic
(1) Blackwell, D.E. (1921-), Oxford, UK
(1) Brueck, H.A. (1905-), Edinburgh, UK
(1) Bruzek, A. (1915-) Freiburg i.Br., German Federal Republic
(1) Bumba, V. (1925-), Ondrejov, Czechoslovakia
(0) De Jager, C. (1921-), Utrecht, The Netherlands
(1) Dollfus, A. (1924-), Meudon, France
(1) Evans, J.W. (1909-), Sacramento Peak, NM, USA
(1) Firor, J.W. Jr. (1927-), Boulder, CO, USA
(1) Fokker, A.D. (1926-), Utrecht, The Netherlands
(3) Friedman, H. (1916-), Washington, DC, USA
(1) Giovanelli, R. (1915-1984), Sydney, Australia
(1) Gnevyshev, M.N.* Kislovodsk, USSR
(1) Goldberg, L. (1913-1987), Cambridge, MA, USA
(1) Houtgast, J. (1908-1982), Utrecht, The Netherlands
(1) Howard, R.F. (1932-), Pasadena, CA, USA
(1) Jaeger, F.W. (1914-), Potsdam, German Democratic Republic
(1) Jefferies, J.T. (1925-), Honolulu, HI, USA
(1) Kiepenheuer, K.O. (1910-1975), Freiburg i.Br., German Federal Republic
(1) Kleczek, J. (1923-), Ondrejov, Czechoslovakia
(1) Krat, V.A. (1911-1983), Poulkovo, USSR
(3) Kundu, M.R. (1930-), Bombay, India
(1) Maxwell, A. (1926-) Cambridge, MA, USA
(1) Michard, R. (1925-), Meudon, France
(1) Mohler, O.C. (1908-1985), Ann Arbor, MI, USA
(3) Newkirk, G.A. Jr. (1928-1985), Boulder, CO, USA
(1) Noyes, J.C. (1920-), Seattle, WA, USA
(2) Noyes, R.W. (1934-), Cambridge, MA, USA
(1) Oehman, Y. (1903-), Saltsjoebaden, Sweden
(3) Parker, E.N. (1927-), Chicago, IL, USA

(1) Pierce, A.K. (1918-), Tucson, AZ, USA
(3) Pikèner, S.B. (1921-1975), Moscow, USSR
(1) Righini, G. (1905-1978), Florence, Italy
(1) Roesch, J. (1915-), Bagneres-de-Bigorre, France
(1) Severny, A.B. (1913-1986), Nauchny, Crimea, USSR
(2) Stepanov, W.E.* Irkutsk, USSR
(1) Suemoto, Z. (1920-), Tokyo, Japan
(1a) Svestka, Z. (1925-) Ondrejov, Czechoslovakia
(1) Takakura, T. (1925-), Mitaka, Japan
(1) Tousey, R. (1908-), Washington, DC, USA
(1) Unno, W. (1925-), Tokyo, Japan
(3) Waldmeier, M.R. (1912-), Zürich, Switzerland
(1) Wild, J.P. (1923-), Sydney, Australia
(1) Zirin, H. (1929-), Pasadena, CA, USA

(0)= the journal's founder; (1)= recruited Mar-June 1966; (1a) = the journal's co-editor; (2) = recruited July-Dec 1966; (3) = elected Oct-Nov 1967; * = birthdate undetermined.

Age Distrib. (1/1/1967): 30-39: 8; 40-49: 17; 50-59: 17; 60-69: 2; Undetermined: 2. *Median Age*: 46.

Country Distrib.: USA: 17; USSR: 5; Czech., France, FRG, Japan, Netherlands: 3; Australia, Sweden, UK: 2; DDR, India, Italy, Switz.: 1.

(*Sources*: C. de Jager to potential ed. board members (16 Mar. 1966), de Jager to ed. board members (13 June 1966), de Jager and Z. Svestka to potential contributors (June 1966), de Jager and Svestka to ed. board members (3 Oct. 1967), all in L. Goldberg Papers, Tucson, AZ; title pages, *Solar Physics* 1 (1967) 3 (1968); and, for birth and death dates, directories and several informants. For a listing of all the board members for the first one hundred volumes, see C. de Jager and Z. Svestka, "Editorial", *Solar Physics* 100 (1985) ix-x.)

COMMENTARY

Elly Dekker *On the paper of Karl Hufbauer**

Karl Hufbauer's thought-provoking study of the development of the branch of astronomy called solar physics into a subdiscipline raises a number of interesting points. A dominating feature of the essay is the tracing of the evolution of a scientific discipline by following the development of a scientific journal exclusively devoted to it. This method is attractive because it applies a more or less quantifiable handle to the complex ways of science history. As such the paper fits well to the theme of this conference, devoted as it is to new trends in the history of science.

However, to what measure is the development of a scientific journal determined by that of the branch of science it serves? Hufbauer considers the journal *Solar Physics.* This has been a great success from its start in 1967 until about 1975. Its annual number of pages and its circulation increased both very rapidly during that period. However, the same has happened to many other science journals in those days, specialized or not, although the former blossomed particularly well. The number of pages accepted by journals of this kind increased on the average by a factor 4 in 10 years, whereas moreover the number of such journals itself more than tripled. Of greater significance might well be the circulation number: with a top circulation of 1500, the journal *Solar Physics* indeed appeared to reach a large part of the astronomical and possibly also the physics community of that time. Certainly, a much larger audience than that of the solar physicists alone was served, which number less than a few hundreds. The interest in the subject solar physics clearly became widely spread through the efforts of *Solar Physics.* In this respect one can indeed speak of a resounding success.

The author attributes this success to a number of factors, of which the birth of space research, initiated by rockets and later re-enforced by the Sputnik program in 1957, is considered the most important one, and solar radio astronomy a good runnerup. I could not agree more with him

*This commentary is partly based on a discussion with H.G. van Bueren, formerly professor in the physical foundations of astrophysics at the University of Utrecht and now chairman of the Netherlands Council for Science Policy.

about the importance of the Sputnik program for the advancement of fundamental next to applied science. The founding of NASA in 1958 provided numerous new opportunities for observational and instrumentally inclined astronomers, and especially for those involved in solar and planetary physics. In a few years' time the budget of NASA rose to a total of over a billion dollars. In 1972 the budget available for space-astronomy amounted to twice that of all groundbased astronomy in the U.S. If indeed the publication explosion in *Solar Physics* can be seen as the direct result of the rapid increase in number of solar physicists (and it has still to be shown that this is so), must not the spectacular success of the Journal be attributed primarily to the Sputnik? And if so, may we conclude that the development of solar physics into a subdiscipline was actually triggered off by the Sputnik program?

The discussion among astronomers about the scientific standards of the journal *Solar Physics* as mentioned by Hufbauer throws an interesting light on the general history of the scientific journal of our times. Various influences as: the publish-or-perish morale, initiatives of the publisher, the commercial pressure for more and extensive publication, the use and misuse of power by referees, the cost problem for institutions (libraries) and individuals; all come to the fore in this discussion. A more extensive comparison between the case of *Solar Physics* and that of the other journal mentioned, also founded by de Jager, namely *Space Science Reviews*, would have been very illuminating in this respect. Superficially the two journals are much alike; intrinsically they differ greatly and their developments varied accordingly.

Most important from the point of view of science history is the question as to the effect of the founding of a specialized journal like *Solar Physics* on the development of astronomy and on Dutch astronomy in particular. More or less in the midst of the high tide of *Solar Physics* another Dutch publication venture brought fruit. On the initiative of Dutch astronomers a number of national (general) astronomical journals in Europe were combined into one European journal, called *Astronomy and Astrophysics*. This is a general astronomical publication (organized into a number of sections) of very high acceptance standards. As such it is directly comparable to the longstanding leading American *Astrophysical Journal*. "A and A" prospered and showed as rapid an expansion as *Solar Physics* did. But, understandably, its section "Sun" never occupied more than at most 10% of the available printing space (and usually but only a few percent). Neither was after 1967 very much attention given to the sun in the *Astrophysical Journal*. This means that the founding of *Solar Physics* has been instrumental in causing a separation between the astrophysicists occupied with the sun and those interested in other astrophysical objects, including stars and nebulae. A similar development has occurred in other subfields of astronomy where specialized journals exist (planetary astrophysics, celestial mechanics). The rift is nowhere complete, but it is unmistakably there. The relevant communities of scientists clearly tend to drift apart, and the smaller one becomes isolated. This is most notable in Holland, where because of its long time concentration on solar physics and directly related subjects, Utrecht observatory has followed quite a different line of development from that of the other two large observatories in Leiden and Groningen. Even in our days, when next to solar physics other astrophysical subjects receive appreciable attention at Utrecht, the rift remains and is even detectable in new developments like radio astronomy. In the early sixties this trend towards isolation was not considered as a danger. In those days solar physics was a field of much more than purely astronomical interest because the Sun was regarded as a laboratory of physical processes taking place under circumstances unattainable at the Earth. Outside scientists working in plasmaphysics and magneto-hydrodynamics therefore showed great interest in solar phenomena. Since about 1975 it has become clear that the aforementioned "solar laboratory" is rather badly operated: many interesting phenomena indeed occur in it, but they are terribly mixed up with each other and therefore lend themselves not very well to accurate analysis followed by more than phenomenological understanding. In the sixties it was the solar flare phenomenon that formed the most important reason for continued study of the sun at all wavelengths and by all sorts of methods. Now, 25 years later, this fundamental phenomenon has not yet been satisfactorily explained. Nevertheless, much has been learned about the sun in the meantime, but the conviction of outside scientists that by observing the sun fundamental physics could be enriched, has been severely damaged. For this reason, the solar physics community is now again more or less on its own - as is shown by the dramatic decrease of papers submitted to *Solar Physics*. The author claims in his article that this decrease should be due to solar physicists publishing more in non-specialized journals. This may be so, and then it indicates another tendency away from the interest in purely solar physics phenomena, but the decreased appeal that solar physics has for other physicists should not be underrated either. Why is it important to realize this? Because it has to do with a central question that occurs throughout the paper (and is not answered in it) and which is of general importance for all science and its history, namely whether there exists a trend of specialized sciences to isolate themselves.

This trend can certainly be recognized in the development of solar physics. But is it correct to conclude, as Hufbauer does, that "the tempering of a separist inclination" mainly came about through the scarcity of resources? Does not rather the trend to move towards more general astrophysics stem from the notion that the field of solar physics is of too limited a scope to support itself in the long run? The growth or decline of a (sub)discipline is not invariably caused by budgetary factors. First and foremost the way in which the relevant research stimulates and is stimulated by other topics and fields of research, and can be applied to wider fields of study, determines its development. It is against this background that the plea of a number of prominent solar physicists to establish a combined effort between solar and stellar astrophysics must be considered. This, then, brings us back to 1937, when Minnaert in his inaugural lecture expounded the following view on the significance of solar physics for general astrophysics: "For Solar Physics not to deteriorate into loose studies of detail, but rather to attain its full and general significance and fulfill its true function, it should be practiced within the framework of Astrophysics in its most general sense."[1]

Now, in 1987, Minnaert's view seems to prevail again, at least in Utrecht. For, in the annual report solar physics is not mentioned as a separate field of research. Instead we read:

> "The solar and stellar studies are complementary: solar observations are used if high spatial, spectral or temporal resolution is required; stellar observations provide a range of stellar mass, angular momentum and evolutionary phase."[2]

Of interest too is the fact that most of those scientists who in my student days worked on solar plasmas and the magneto-hydrodynamics of the sun now are otherwise engaged:

> "The group studies magneto-hydrodynamics and plasma processes in highly energetic astrophysical plasmas in accretion disks around single stars, in binaries and under extragalactic conditions."[3]

If the developments at the Utrecht observatory are typical we must conclude that the development of solar physics into a separate subdiscipline was rather short lived. This being the case, I like to conclude this commentary with the remark that Karl Hufbauer's study of 'Solar Physics evolution into a subdiscipline (1945-1975)' is a fine example of contemporary science history, but one that needs extending up to the present day.

National Museum of the History of Science and Medicine (Leiden)

1. M.G.J. Minnaert, *De betekenis der zonnephysica voor de astrophysica*. Inaugural lecture, 18 Oct 1937. (Groningen, 1937) 16.
2. *Annual Report of Utrecht Observatory* (Utrecht, 1985) 5.
3. *Annual Report*, 2.

Martin J. Klein

BOLDNESS AND CAUTION: ASPECTS OF NIELS BOHR'S SCIENTIFIC STYLE

In 1922 Albert Einstein proposed to his fellow academicians in Berlin that they elect Niels Bohr as a corresponding member. Less than a decade had passed since Bohr launched his attack on the problems of atomic physics and chemistry, but his ideas already so dominated the field that, as Einstein remarked, "all earlier attempts at explanation seem to come from an age long since vanished". Einstein did not limit his praise of Bohr to the positive achievements which had made him "one of the greatest scientific discoverers of our time". He had also been struck by the remarkable way in which Bohr approached physics:

> "What is so wonderfully appealing about Bohr as a scientist is a rare blend of boldness and careful consideration; seldom has a scientist been endowed in such measure as he with an intuitive grasp of hidden things and with a keen critical sense. With all his knowledge of details his gaze is steadily directed to the underlying principles".[1]

No physicist would have argued against Einstein's characterization of Bohr as a bold thinker. In 1913, when his papers "On the Constitution of Atoms and Molecules"[2] appeared, Bohr's boldness had shocked some of his contemporaries into extreme reactions. "Bohr's work on the quantum theory of the Balmer formula ... has driven me to despair", Paul Ehrenfest wrote to H.A. Lorentz. "If this is the way to reach the goal, I

1. A. Einstein, "Wahlvorschlag für N. Bohr zur Aufnahme als korrespondierendes Mitglied in die Akademie d. Wiss.", in: Christa Kirsten and Hans-Jürgen Treder eds., *Albert Einstein in Berlin 1913-1933*, Teil I. *Darstellung und Dokumente* (Berlin, 1979) 130.
2. N. Bohr, "On the Constitution of Atoms and Molecules", *Philosophical Magazine* 26 (1913) 1-25; 476-502; 857-875. Reprinted with a valuable historical introduction by Léon Rosenfeld: N. Bohr, *On the Constitution of Atoms and Molecules* (Copenhagen and New York, 1963). Reprinted again in: Ulrich Hoyer ed., *Niels Bohr, Collected Works*, Vol. 2, *Work on Atomic Physics (1919-1917)* (Amsterdam and New York, 1981) 159-233. Bohr's *Collected Works* were under the General Editorship of Léon Rosenfeld for Volumes 1-3 and since Volume 5, Erik Rüdinger. Publication of the series began in 1972. I shall refer to the series simply as *Works*.

must give up doing physics".[3] Otto Stern and Max von Laue came to the same decision for the same reasons[4], while Erwin Schrödinger described one of Bohr's major innovations as "*monstrous*" and "almost *inconceivable*".[5] None of these young men could be called timid, yet they were all provoked into these hyperboles of horror by what Bohr had done. Despite the evident and manifold success of this work, these readers rejected his approach. All of them would surely have been ready to adopt a phrase of Robert A. Millikan and describe Bohr as "bold, not to say reckless".[6]

Yet Bohr was anything but reckless. He had not really accepted the advice Odysseus offered to Neoptolemus: "But what a glorious prize is victory! / Therefore be bold; we will be just hereafter".[7] To see the "careful consideration" that was blended with the boldness of Bohr's ideas did require some sympathy with his way of thinking and with his view of the current state of physics. Bohr had recognized that physicists had really entered an unknown and uncharted domain in their efforts to understand matter at the atomic level, a domain in which one could no longer rely on the validity of either mechanics or electrodynamics. With no established paths to follow, one had to be as sensitive as a frontier scout to the hints offered by the surrounding terrain itself, and trust one's own sense of what was essential. Those who could not give up the intellectual security that comes with working within the clear structure of an established theory could hardly be expected to appreciate Bohr's careful advance into the unknown.

No one was in a better position to understand Bohr's way of proceeding than Einstein. At the first Solvay Conference in 1911 he had emphasized the inadequacy of classical physics for providing a theoretical understanding, while recognizing that "the theory of quanta in its present form ... does not really constitute a theory in the ordinary sense of the word". It was surely not yet "a theory that could be developed in a coherent manner". Einstein was therefore trying to see "which of the general principles of physics we can rely on for solving the problems that

3. P. Ehrenfest to H.A. Lorentz, 25 August 1913.

4. See Abraham Pais, *Inward Bound. Of Matter and Forces in the Physical World* (Oxford and New York, 1986) 208.

5. E. Schrödinger to H.A. Lorentz, 6 June 1926, in: K. Przibram ed., transl. M.J. Klein, *Letters on Wave Mechanics* (New York, 1967) 61.

6. R.A. Millikan, "A Direct Photoelectric Determination of Planck's 'h' ", *Physical Review*, ser. 2, 7 (1916) 355.

7. Sophocles, *Philoctetes* (transl. Thomas Francklin), in: Whitney J. Oates and Eugene O'Neill, Jr, eds., *The Complete Greek Drama*, Vol. I (New York, 1938) 558.

confront us".[8] When he looked back on this period thirty-five years later in his "Autobiographical Notes", Einstein remembered how he had felt about the state of theoretical physics. "It was as if the ground had been pulled out from under one, with no firm foundation to be seen anywhere upon which one could have built". And it is surely significant that his next sentences were:

> "That this insecure and contradictory foundation was sufficient to enable a man of Bohr's unique instinct and sensitivity to discover the principal laws of the spectral lines and of the electron shells of the atoms, together with their significance for chemistry, appeared to me as a miracle — and appears to me a miracle even today. This is the highest form of musicality in the sphere of thought".[9]

Bohr had not done all of this in 1913, but even then Einstein saw his successful work on the hydrogen spectrum as "an enormous achievement" and recognized it as "one of the greatest discoveries".[10]

Einstein was obviously not the only one to appreciate the remarkable features of Bohr's work. I am concerned here, however, not with the range of reactions to "the great trilogy" of 1913, but rather with Bohr's own thoughts on the subject. How did he view those papers that had suddenly become the focus for discussion by physicists all over Europe? What features of his work seemed most solid to its author? The papers presented a mix of very general, startlingly original and, as it turned out, extraordinarily fruitful ideas with various rather special assumptions and models, and even some concepts left over from an earlier and contradictory approach to the emission of spectra. How far did Bohr see into this mixture after its publication? These questions are not easily answered, but Bohr's papers, manuscripts, and correspondence do throw some light on them. One particularly interesting source is the occasional correspondence between Bohr and a friend whose name is not so widely known — the Swedish physicist Carl Wilhelm Oseen. It seems that for several years, at least, Oseen was one of those to whom Bohr could express his ideas about physics most freely and openly, securely confident of receiving a sympathetic and encouraging response.[11]

8. A. Einstein, "Discussion du rapport de M. Einstein", in: P. Langevin and M. de Broglie eds., *La Théorie du Rayonnement et les Quanta* (Paris, 1912) 436.

9. A. Einstein, "Autobiographical Notes", in: Paul Arthur Schilpp ed., *Albert Einstein - Philosopher Scientist* (Evanston, Illinois, 1949) 2-95, quotation from pp. 45-47.

10. George Hevesy to N. Bohr, 23 September 1913, and G. Hevesy to E. Rutherford, 14 October 1913. Quoted in Rosenfeld, *Bohr, Constitution,* xlii. The term "great trilogy" for Bohr's 1913 papers comes from Rosenfeld's introduction in this reference.

11. Carl Wilhelm Oseen (1879-1944) became Professor of Mechanics and

Oseen was six years older than Bohr, and had been Professor of Mechanics and Mathematical Physics at the University of Uppsala since 1909. When Bohr completed his doctoral dissertation, "Studies on the Electron Theory of Metals", in April 1911, one of the copies he sent out went to Oseen, whom he had not yet met.[12] Oseen actually studied the dissertation before writing to Bohr to thank him. In addition to expressing his "pleasure that such a solid work of mathematical physics has appeared in one of our Nordic countries", Oseen commented on only one aspect of Bohr's work.

> "I wish especially to express my pleasure over the criticism to which you have subjected the flood of papers on the subject in question. Partly with this particular literature in mind, I have long been of the opinion that what mathematical physics is most in need of at present is criticism".[13]

And indeed Bohr's critical abilities were much in evidence in his thesis as he carefully analyzed and often took issues with the work of J.J. Thomson, Ludwig Boltzmann, H.A. Lorentz, James Jeans, Peter Debye, and others. His bold conclusion that free electrons in a metal cannot give rise to magnetic properties, if classical physics is correct, may or may not have been noted by Oseen, but Bohr must have been gratified to see that his effective criticism had made its mark.

Oseen was in Copenhagen during the summer of 1911 for the Scandinavian Mathematical Congress where he met and became friends with Niels Bohr and his younger brother Harald, already widely known as both mathematician and Olympic soccer star.[14] In December of that year Bohr wrote to Oseen from Cambridge.[15] His twelve page letter is full of Bohr's lively responses to the great variety of new people, institutions and ideas in his life. He writes of the Cavendish Laboratory "where there exists a lack of order that you cannot imagine", and J.J. Thomson "who is unbelievably full of ideas and a very clever man", but who "has so little time" that he had not yet read the English translation of the thesis Bohr was so anxious to discuss with him. Bohr was trying to do both

Mathematical Physics at the University of Uppsala in 1909, a position he held until 1933, when he became Director of the Nobel Institute. See J. Rud Nielsen ed., *N. Bohr, Collected Works*, Vol. I, *Early Work (1905-1911)* (Amsterdam and New York, 1972) 101.

12. N. Bohr, "Studies on the Electron Theory of Metals", in: *Works*, Vol. I, 291-395.
13. C.W. Oseen to N. Bohr, 5 July 1911. *Works*, Vol. I, 101-102.
14. L. Rosenfeld and E. Rüdinger, "The Decisive Years 1911-1918", in: S. Rozental ed., *Niels Bohr. His Life and Work as Seen by His Friends and Colleagues* (Amsterdam, 1967) 39.
15. N. Bohr to C.W. Oseen, 1 December 1911, *Works*, 1, 426-431. Also N. Bohr to Ellen Adler Bohr, 6 December 1911, *ibid.*, 535.

experimental and theoretical work at the same time, and finding it "not so easy to do". He comments freely on many aspects of the electron theory of metals, praising a recent paper by S.B. McLaren but finding a paper by J.W. Nicholson to be "perfectly crazy". Bohr had started to study S.H. Burbury's book on the kinetic theory of gases, and explains to Oseen why some of Burbury's results are wrong.

This letter also includes a couple of paragraphs in which Bohr speculates about the meaning of Boltzmann's H-theorem and its relationship to the second law of thermodynamics. He breaks off, however, by remarking parenthetically: "I dare not express myself too definitely tonight, and it is the kind of problem that almost always slips through our fingers at the very moment we think that we have really grasped it", a sentiment that many physicists might be inclined to echo. And Bohr concluded by telling Oseen that he is "at the moment ... very enthusiastic about the quantum theory", particularly its experimental aspects, and that he is "looking forward immensely to really getting into all of it next term".

I have described and quoted from this letter at some length to show how openly Bohr expressed his opinions on physics and physicists to Oseen. Letters cannot always be taken at their face value: a letter is addressed to a particular individual and its style and content are chosen accordingly. Bohr was, however, evidently comfortable with Oseen, and enjoyed having a friend who knew physics and seemed to share his outlook and to whom he could state his ideas, however preliminary they might be. That he could write to Oseen in his native Danish — no barrier to a Swedish reader — at a time when virtually all his other scientific correspondence had to be carried on in English or German, must have added to Bohr's sense of comfort. Their exchanges, though not very frequent, are always interesting.

Oseen did not react to Bohr's work on the hydrogen spectrum until November 1913. By this time Bohr was aware of at least some of the different ways in which physicists responded to his work. Harald had informed him that the Göttingen group, and especially its youngest members like Max Born and Erwin Madelung found his assumptions "too 'bold' and 'fantastic' ".[16] This opinion preceded the experimental confirmation of Bohr's explanation of the Pickering-Fowler lines, the result that led Einstein to the enthusiastic response already quoted. At the British Association meeting at Birmingham in September, Bohr had heard his work praised by Jeans as "a most ingenious and suggestive, and

16. H. Bohr to N. Bohr, n.d. (Fall 1913), *ibid.,* 567.

UNIV DUNELM

I think we must add convincing, explanation of the laws of spectral series".[17]

And now here was Oseen's letter with the most gratifying words of all.[18]

> "What I would like to tell you first is that, although I already knew the direction of your thinking as well as some of its results, I was still surprised at one point by the beauty of your result. This was the connection between h and the Balmer-Rydberg constant. As far as one can see, on this point you have gone beyond the region of hypotheses and theories and into that of truth itself. Higher no theorist can reach, and I congratulate you with all my heart".

This praise came from a man well aware that Bohr's work raised new questions. He wondered "how the Maxwell-Lorentz theory should be modified to allow for the existence of an atom" like Bohr's, saying that "every time I see more clearly how deep the modification must reach!"

In December 1913 Bohr was invited to lecture to the Danish Physical Society at Copenhagen. The invitation gave him an opportunity to rethink the arguments he had worked out and written up so quickly earlier that year. He was to talk about the hydrogen spectrum, the subject of the first and most successful part of his trilogy. Although Bohr had been working on atomic structure since June 1912, the problem of spectra was not included in his considerations until much later.[19] As late as January 31, 1913 he wrote to Rutherford: "I do not at all deal with the question of calculation of the frequencies corresponding to the lines in the visible spectrum".[20] By March 6th, however, he had completed a draft of Part I of his series "On the Constitution of Atoms and Molecules" dealing with just that question, and its final version was finished only a month later, since it bears the date April 5, 1913.[21] Bohr must therefore have welcomed the chance to review the subject and put his thoughts in better order for presentation to the Physical Society.

We know that he worked hard at his manuscript. "I am very busy these days with a lecture 'On the hydrogen spectrum' that I shall deliver Friday in the Physical Society", he wrote to Harald a week before the date.[22] The lecture itself shows that this time, at least, Bohr's efforts to achieve clarity

17. *Ibid.*, 124.
18. C.W. Oseen to N. Bohr, 11 November 1913, *Works*, Vol. 2, 552-553.
19. See Rosenfeld's and Hoyer's introductions in the references of note 2. Also see the important study by John L. Heilbron and Thomas S. Kuhn, "The Genesis of the Bohr Atom", *Historical Studies in the Physical Science* 1 (1969) 211-290.
20. N. Bohr to E. Rutherford, 31 January 1913, *Works*, Vol. 2, 579-580.
21. N. Bohr to E. Rutherford, 6 March 1913, *ibid.,* 581-583. See also p.185.
22. N. Bohr to H. Bohr, 13 December 1913, *Works,* Vol. I, 567.

were successful.[23] Its text is particularly interesting for a number of reasons, but I shall restrict my comments to what it tells us about the state of Bohr's thinking some months after the original publication of his work.

He began by pointing out the "beautiful and simple laws concerning the line spectra of the elements", those laws that he had not known — or at least had not incorporated into his own work — less than a year earlier. Despite their simplicity none of the theories previously proposed to account for them had offered anything like a plausible explanation of the spectral laws; this was not surprising since existing theoretical structures were known to be inadequate at the atomic level. "You will understand", Bohr then said,

> "that I shall not attempt to propose an explanation of the spectral laws; on the contrary I shall try to indicate a way in which it appears possible to bring the spectral laws into close connection with other properties of the elements, which appear to be equally inexplicable on the basis of the present state of the science".[24]

This remarkable statement set the tone for what was to follow. Bohr next sketched the two principal sources of his own work. First was the quantum of energy introduced by Planck. Despite the inconsistencies in Planck's work and regardless of how theory might eventually progress, it was clear that energy quanta must play a role "in investigations of the properties of atoms and particularly in connection with any explanation of the spectral laws in which such phenomena as the emission and absorption of electromagnetic radiation are concerned".[25] The second ingredient was Rutherford's nuclear atom, whose positive features did not hide the fact that it could not be discussed on the basis of "the ordinary electrodynamics". This, according to Bohr, was not a defect.

> "The fact that the deficiencies of the atomic model we are considering stand out so plainly is therefore perhaps no serious drawback; even though the defects of other atomic models are much better concealed they must nevertheless be present and will be just as serious".[26]

Bohr was trying to emphasize that the present state of atomic physics did not provide a basis for theoretical explanations of a familiar and acceptable type.

> "We stand here almost entirely on virgin ground, and upon introducing new assumptions we need only take care not to get into contradiction with experiment.

23. N. Bohr, "On the Spectrum of Hydrogen", Address delivered before the Physical Society in Copenhagen, 20 December 1913, *Works*, Vol. 2, 281-301.
24. *Ibid.,* 285-286.
25. *Ibid.,* 289.
26. *Ibid.,* 291-292.

> Time will have to show to what extent this can be avoided; but the safest way is, of course, to make as few assumptions as possible."[27]

Note that it was contradiction with experiment that was to be avoided, but that no such condition could be made about contradictions with previously held theories.

The crucial assumption introduced energy quanta into the discussion of the emission (and absorption) of light by atoms. If the spectrum of an element contains a spectral line of frequency ν, then according to Bohr its atoms can emit energy in the amount $h\nu$, where h is Planck's constant. The energy of the atom before and after emitting this radiation can be denoted by E_1 and E_2 respectively, where these are related by the equation,

$$(1) \qquad h\nu = E_1 - E_2 \,.$$

The states of the atom before and after the emission occurs Bohr called stationary states, and he remarked that the terminology simply indicates "that they form some kind of waiting places between which occurs the emission of the energy corresponding to the various spectral lines".[28]

Now the most general of the "beautiful and simple laws" describing the line spectra of the elements, the Rydberg-Ritz combination principle, expressed all spectral frequences as the differences of "term values". Bohr's basic equation (1) thus gave an immediate interpretation of this principle with the Rydberg-Ritz "term values" proportional to the energies of the stationary states.

Bohr then applied this idea to the line spectrum of atomic hydrogen. Using the Balmer formula for the term values of the lines he was led at once to the energy E_n of the atom in state n,

$$(2) \qquad E_n = -(Rhc/n^2) \,,$$

where c is the velocity of light, R is the empirical Rydberg constant, and n is an integer. One could also calculate the frequency ω_n with which an electron will rotate in a circular orbit about a fixed nucleus with an equal and opposite charge. Taking the mass of the electron to be m and its charge -e, this frequency is readily found to be given by the equation,

27. *Ibid.*, 292.
28. *Ibid.*, 293.

(3) $$\omega_n^2 = \frac{2|E_n|^3}{\pi^2 e^4 m} ,$$

where E_n is the total energy of the revolving electron. This result assumes that the motion in this stationary state can be described by "the ordinary mechanics", since as Bohr pointed out "there is, for the present at least, no other means available".[29] Nevertheless one could not expect that the frequency of the light emitted in a transition from one stationary state to another, as given by equation (1), had any simple relation to the frequencies with which the electron revolved about the nucleus in these two states. That light of a particular frequency could be emitted by an atom when nothing in the atom had a motion of the same frequency had been a stumbling block for Bohr's critics. It was just this frequency discrepancy that Schrödinger called "monstrous", and "almost *inconceivable*".

Bohr granted this point, in effect, as he immediately remarked to his audience: "You understand, of course, that I am by no means trying to give what might ordinarily be described as an explanation; nothing has been said here about how or why the radiation is emitted".[30] But even if one could not expect a general connection between the frequencies of the electron's revolutions and the frequencies of the emitted spectral lines, as classical theory would have required, there was a link between them. Classical electrodynamics could be expected to give valid results when used to calculate the emission of low frequency lines. Bohr justified this claim by reminding his audience that in 1903 Lorentz had published a successful classical calculation of the low frequency end of the blackbody radiation spectrum, finding a result in agreement with the appropriate limit of Planck's distribution law.[31]

Since equations (2) and (3) show that low frequencies are equivalent to large values of n, Bohr looked at the frequency emitted in the transition from state (n + 1) to state n when n is very large, and obtained the result,

(4) $$\nu = (2Rc/n^3) ,$$

29. *Ibid.,* 294.

30. *Ibid.,* 294-295.

31. H.A. Lorentz, "On the Emission and Absorption by Metals of Rays of Heat of Great Wave-Lengths", *Proceedings of the Section of Sciences of the Koninklijke Akademie van Wetenschappen* 5 (1903) 666-685. Bohr probably quoted this work rather than Rayleigh's since he had studied Lorentz carefully while preparing his dissertation. See *Works*, Vol. 1, 357-379.

with the help of equations (1) and (2).[32] When this is equated to ω_n and the result is simplified, one obtains an expression for the Rydberg constant in terms of the fundamental atomic constants,

$$(5) \qquad R = \frac{2\pi^2 me^4}{ch^3} \; .$$

The measured value of R agreed with the value calculated from this equation within a percent or so, an agreement "as good as could be expected, considering the uncertainty in the experimental determination of the constants e, m, and h".[33]

The argument Bohr had used gave no direct insight into the stationary states for values of n that were not large.[34] One could combine the equations given above to show that, for any n, the energy and rotation frequency were related by the equation,

$$(6) \qquad |E_n| = \tfrac{1}{2} nh\omega_n \; ,$$

a result "analogous to Planck's assumption concerning the energy of a resonator". In this lecture, however, Bohr departed sharply from his approach in the original paper, saying: "When we consider how differently the equation is employed here and in Planck's theory it appears to me misleading to use this analogy as a foundation, and ... I have tried to free myself as much as possible from it". That analogy had "originally led [Bohr] into these considerations", but now it was time to put it aside.[35]

There is no need to follow the rest of Bohr's lecture here, but I do want to quote from his last two sentences. He hoped that his talk had been clear enough for his audience to "have appreciated the extent to which these considerations conflict with the admirably coherent group of conceptions which have been rightly termed the classical theory of

32. Bohr had used a somewhat different version of this argument in his original paper. See *Works*, Vol. 2, 172-174. Also see Heilbron and Kuhn, "Genesis", 273-277.

33. Bohr, "On the Spectrum", 295.

34. As Bohr formulated it in his original paper, the essential assumption is "that the frequency of the radiation emitted during the passing of the system between successive stationary states will coincide with the frequency of revolution of the electron in the region of slow vibrations". *Works*, Vol. 2, 174. For further discussion of the correspondence principle see Klaus Michael Meyer-Abich, *Korrespondenz, Individualität und Komplementarität* (Wiesbaden, 1965) 34-41; 46-101.

35. Bohr, "On the Spectrum," 295-296.

electrodynamics". "On the other hand", he concluded, "I have tried to convey to you the impression that just by emphasizing this conflict, it may be also possible in the course of time to discover a certain coherence in the new ideas".[36]

We know how Bohr thought about this lecture shortly after he gave it from his comments in a letter to Oseen written early in March 1914.

> "Without changing my point of view in any essential way, I have tried to make my ideas somewhat clearer by changing the order of my arguments. The train of thought I have been following can be seen in the small paper which I enclose (it is only a semi-popular talk I gave just before Christmas in the Physical Society in Copenhagen.) By following exactly the train of thought which is given there it is possible in any case to say something about the Stark effect. I have written a paper on this and on the Zeeman effect, which I hope will appear in Phil. Mag. in a few days".[37]

That paper did indeed begin with just the "train of thought" set forth in Bohr's lecture.[38]

He placed a high value on this lecture throughout his life. In 1922 he grouped it with two other papers in a book, *The Theory of Spectra and Atomic Constitution* that also appeared in German and French versions.[39] His associates Léon Rosenfeld and Erik Rüdinger report that Bohr always considered it "one of his best and clearest lectures".[40]

The most remarkable evidence for Bohr's opinion of this work comes from Thomas Kuhn's recent account of his discussions with Bohr in the fall of 1962, only weeks before Bohr's death.[41] Kuhn was trying to get Bohr to comment on his thinking about the connections between the Rutherford atom and the quantum in the period before the Balmer formula had been drawn to his attention. Bohr rejected the idea that he could have been thinking along these lines in the absence of the crucial

36. *Ibid.*, 301. I have used the slightly revised translation given by Rosenfeld and Rüdinger, "Decisive Years", 60-61.

37. N. Bohr to C.W. Oseen, 3 March 1914, *Works*, Vol. 2, 555-556.

38. N. Bohr, "On the Effect of Electric and Magnetic Fields on Spectral Lines", *Philosophical Magazine* 27 (1914) 506-524; *Works* Vol. 2, 349-368.

39. N. Bohr, *The Theory of Spectra and Atomic Constitution* (Cambridge, 1922); *Drei Aufsätze über Spektren und Atombau* (Braunschweig, 1922); *Les Spectres et la Structure de l'Atome* (Paris, 1923).

40. Rosenfeld and Rüdinger, "Decisive Years", 60-61.

41. T.S. Kuhn, "Revisiting Planck", *Historical Studies in the Physical Sciences* 14 (1984) 231-252. Quotation on pp. 247-248. These discussions took place before the unpublished "Rutherford Memorandum" had come to light, but Kuhn had already surmised that Bohr's first attempt to apply quantum concepts to the Rutherford atom had dealt only with the ground state. (See the references in note 19.) The last section of Kuhn's paper is an important contribution to the historiography of science.

clue provided by spectra. Kuhn reports that Bohr's assistant quoted Bohr as saying that "he should not have published so soon". He was referring to the great trilogy, for he went on: "Parts II and III should have been entirely suppressed and Part I should have been withheld until recast in the form of the lecture delivered in Copenhagen in December". Kuhn also remembers Bohr showing him his own reprint of the Christmas lecture and asking, "It's all right there, isn't it?" This came after Bohr had "muttered half to himself, 'Perhaps it was a mistake to put the paper into print so fast. Perhaps I should have waited until I had it right'". Historians may shudder at these words, but one can understand Bohr's feelings.

In July 1914 Niels and Harald Bohr took a trip to Germany where Niels met several well-known physicists and talked in the seminars at Göttingen and Munich.[42] "I had never met any German physicists before, and had much pleasure in talking with them", he reported to Oseen in September.[43] The Bohr brothers were in Germany when the war began, but, as Niels wrote, "We just managed to return before things became difficult for foreigners". As usual his letter to Oseen helps to illuminate his current thinking.

Despite the brilliant confirmation of his account of the Pickering-Fowler spectrum by Evans's experiments, Bohr had no illusions that further progress would be easy. He saw no way of dealing with systems containing more than two particles. "I tend to believe that in this problem there are buried very considerable difficulties", he wrote, "which can only be avoided by departing from the usual considerations to an even greater extent than has been necessary up to now, and that the preliminary success is only due to the simplicity of the systems considered". Bohr hoped that experiment might help: the relation between frequency and energy difference expressed in equation (1) could be tested "by finding the energy which must be supplied to a system in order to make it emit a certain spectral line", and such experiments were already in progress in several laboratories.

Bohr described to Oseen the conversations he had had with Peter Debye in Göttingen about the basis of his theory.

> "For the most part he took a friendly line, but stated that if there should be any reality in this kind of considerations, there had to be in his opinion a general principle which allowed one to understand the connection between the quantum theory and the usual electrodynamics".

42. Rosenfeld and Rüdinger, "Decisive Years", 63-64.
43. N. Bohr to C.W. Oseen, 28 September 1914, *Works*, Vol. 2, 560-563.

This did not suit Bohr at all.

> "I tried to say that the necessity of such a principle was perhaps not evident, and that the problem which classical mechanics and electrodynamics had tried to solve perhaps was very different from the one which the phenomena confronted us with, that the possibility of a comprehensive picture should perhaps not be sought in the generality of the points of view, but rather in the strictest possible limitation of the applicability of the points of view".

This was a theme Bohr would develop more fully over a decade later in his principle of complementarity, but in 1914 he was still far from certain about the very basis of his work. "I long so much for and need so much to really talk to you about all these matters, and to really learn whether you can accept the direction which my thoughts have taken, or if it seems to you superficial and premature", he wrote to Oseen, and then confessed to his own recurrent "doubts about the whole thing".

Whatever hesitations Bohr did have, he was quite sure that the "experimental reality" of the quantum theory could "hardly be doubted", as he wrote to Oseen a year or so later.[44] By this time he could compare his own approach with that taken by two accomplished mathematical physicists whom he had already met, Debye and Sommerfeld. The comparison was reassuring to Bohr. In their attempts at a theory of dispersion[45] they had ignored the most important lesson of his papers: the characteristic frequencies that determine dispersion must follow "the laws for quantum emission", and so "dispersion, whatever its explanation, cannot be calculated from the motion of electrons and the usual electrodynamics, which does not have the slightest connection with the frequencies considered". Debye and Sommerfeld had not yet seen the need to break with the old ways of thinking. Despite his criticism Bohr wrote that he was

> "so happy that two such eminent mathematicians have started taking an interest in the constitution of atoms and molecules; if it is finally possible to make more advances along the well-known routes, they surely will be able soon to reach far".

This double-edged remark of Bohr's proved prophetic.

Only three months later Bohr was writing to Sommerfeld to thank him for his "most interesting and beautiful papers", the papers in which Sommerfeld explained the fine structure of the lines in the hydrogen spectrum by treating the electron's motion relativistically.[46] Different as

44. N. Bohr to C.W. Oseen, 20 December 1915, *Works*, Vol. 2, 566-568.

45. For a brief discussion and references to the papers by Debye and Sommerfeld and to a critical review of their work by Oseen, see *Works*, Vol. 2, 336-340.

46. N. Bohr to A. Sommerfeld, 19 March 1916, *Works*, Vol. 2, 603-604. This is the draft written in English by Bohr.

Sommerfeld's elegant mathematical methods were from Bohr's own, he welcomed the progress they brought. "I do not think that I have ever enjoyed the reading of anything more than I enjoyed the study of them", Bohr wrote.

At the moment Sommerfeld's work reached him — at Manchester, where he was Reader in Physics from the autumn of 1914 until July 1916 — Bohr had a paper in the course of publication at the *Philosophical Magazine*. Although he had already read the proofs of his paper, "On the Application of the Quantum Theory to Periodic Systems", and it was scheduled to appear in the issue of April, 1916, he immediately withdrew it from the journal, postponing its publication indefinitely.[47] Bohr's article was an attempt to show that, at least for periodic systems, one could formulate the postulates of the quantum theory in "a mutually consistent form which covers the various very different applications". He had been satisfied with what he had done, but when he read Sommerfeld's new work he realized that it had "quite changed the present state of the quantum theory",[48] and that he would have to "consider it all again in view of all, for which your papers have opened my eyes".[46]

Bohr felt this need to "consider it all again" repeatedly during the years when the quantum theory seemed so questionable, when progress was anything but uniform. His great monograph "On the Quantum Theory of Line Spectra"[49] represented the first stage of rethinking after the aborted paper of 1916. It too was described as "an attempt to discuss the different applications from a uniform point of view, and especially to consider the underlying assumptions in their relations to ordinary mechanics and electrodynamics". Bohr hoped "to throw some light on the outstanding difficulties by trying to trace the analogy between the quantum theory and the ordinary theory of radiation as closely as possible".[50] This monograph, in which the correspondence principle was developed far beyond the original germ of the idea which we saw in Bohr's 1913 work, was to have comprised four parts. Part One appeared in March 1918, and Part Two, revised to incorporate the latest developments, came out in December of that year. The third part proved to be intractable when

47. See *Works*, Vol. 2, 341-342. The paper is reproduced from the proof sheets on pp. 431-461 of the same volume.

48. N. Bohr to C.W. Oseen 17 March 1916, *Works*, Vol. 2, 571-573.

49. N. Bohr, "On the Quantum Theory of Line-Spectra", *Det Kongelige Danske Videnskabernes Skrifter, Naturvidenskabelig og Mathematisk Afdeling, 8 Raekke*, IV (1918-1922). J. Rud Nielsen ed., *N. Bohr, Collected Works*, Vol. 3, *The Correspondence Principle (1918-1923)* (Amsterdam and New York, 1976) 65-184.

50. *Ibid.*, 70.

Bohr tried to recast it as his ideas changed, and in 1922 he finally published what he had had ready in 1918 with a few pages of comment on later developments. The last part was dropped entirely, since Bohr planned "a new series of papers" setting forth his current ideas.[51]

The new series was to have been the outcome of Bohr's efforts in 1921 and 1922 to work out a comprehensive treatment of the quantum theory as he saw it. He was not at all satisfied with the way any of the books, including Sommerfeld's, dealt with the principles, and felt that they did not do "justice to the actual content of the theory".[52] He was feeling "scientifically lonesome" as he wrote to Sommerfeld in the spring of 1922 and thought that his work on the systematic treatment of the principles had "been received with very little understanding". Bohr's next sentences are particularly striking.

> "For me it is not a matter of didactic trifles but of a serious attempt to achieve such an inner coherence that one can obtain a secure basis for the further development. I understand quite well how little the matters are clarified as yet, and how helpless I am at expressing my thoughts in easily accessible form ... ".[53]

The new series, "On the Application of the Quantum Theory to Atomic Structure", began bravely with "Part I: The Fundamental Postulates" which appeared in German in 1923 and in English translation the following year.[54] The stated purpose of this paper was much like that of Bohr's earlier attempts to set forth a coherent picture of the current state of the quantum theory, and its contents accurately reflected the progress that had been made and the difficulties that still seemed insurmountable. As Pauli once remarked, the shades of meaning that Bohr tried to convey by his reformulation of the postulates underlying the theory exemplify the subtlety of his style. They cannot be appreciated without examining the successive versions that appeared in his various papers.[55] A second paper in the series, dealing with "The Theory of the Series Spectra", was also completed in 1923, but it never went into print. The methods available at that time could not cope with such thorny

51. *Ibid.*, 167.
52. N. Bohr to Arthur Haas, 11 April 1922, *Works*, Vol. 3, 40.
53. N. Bohr to A. Sommerfeld, 30 April 1922, *Works*, Vol. 3, 39.
54. N. Bohr, "Über die Anwendung der Quantentheorie auf den Atombau. I. Die Grundpostulate der Quantentheorie", *Zeitschrift für Physik* 13 (1923) 117. English transl. by L.F. Curtiss, "On the Application of the Quantum Theory to Atomic Structure", *Proceedings of the Cambridge Philosophical Society, Supplement* (1924); *Works*, Vol. 3, 455-499.
55. W. Pauli, "Niels Bohr on His 60th Birthday", *Reviews of Modern Physics* 17 (1945) 97-101. Quotation on p. 99.

problems as the anomalous Zeeman effect and the helium atom, so the paper was put aside and the series came to an abrupt end.[56]

Several years earlier Bohr had found an occasion to express some of his attitudes to the way physics was going. In July 1919 he had received a letter from Charles G. Darwin, his friend since Bohr's earliest days in Manchester.[57] Stimulated by an earlier letter of Bohr's, Darwin had written out his general views of the problems of the quantum theory and about the dubious status of the conservation laws in particular. "I have long felt that the fundamental basis of physics is in a desperate state", Darwin began the manuscript that accompanied his letter to Bohr. He went on to emphasize the contradictions on which the quantum theory seemed to rest, and hoped that he could "knock away the props of classical physics one by one" until the "difficulties have become reconciled".

Bohr was moved to an immediate response and began to draft a letter to Darwin with the help of his student Oskar Klein.[58] "All progress in science emphasizes difficulties", Bohr began. "All progress in physics no proofs whatever but only simple connections of different conceptions". (Bohr had been thinking about his own work this way for some years. There is a manuscript page from 1914 on which he wrote "Deduction of Rydberg's Konstant is not a *deduction* in the ordinary sense, but a necessary *condition* for the connection between the Quantum theory and the ordinary ideas of radiation".)[59]

Most interesting in Bohr's draft letter to Darwin are his general remarks.

> "I appreciate most strongly your critical attitude against the usually accepted proofs of the quantum theory and for my own part I take quite a similar attitude to them without however having tried to concentrate the criticism on any distinct point of the deductions. I look on most general reasoning in science as opportunistic (success- or unsuccessful) relations between conceptions more or less defined by other conceptions and helping us to overlook things and from that respect it can hardly be denied that the introduction of the quantum idea has helped tremendously to obtain rational relations between the large classes of phenomena as well of statistical character as of non-statistical".[58]

56. See *Works*, Vol. 3, 41-44. The German text and an English translation of this manuscript are on pp. 502-531 and 532-558 of the same volume.

57. C.G. Darwin to N. Bohr, 20 July 1919, Klaus Stolzenburg ed., *N. Bohr, Collected Works*, Vol. 5, *The Emergence of Quantum Mechanics (Mainly 1924-1926)* (Amsterdam and New York, 1984) 314-315. See also pp. 13-19.

58. N. Bohr to C.G. Darwin, July 1919 draft, *Works*, Vol. 5, 15-16. The drafts in this and the following note were written in English.

The draft goes on to deal with problems concerning energy conservation and quanta, but it breaks off without coming to any conclusions, and no answer to Darwin's letter was sent at this time. As Bohr put it when he finally did write to Darwin some two and a half years later, "... but due to the inherent difficulties in the matter I did not succeed that time to bring my views in satisfactory form, and due to other pressing work I put the concluding of the letter off with the familiar result that it was never finished". Bohr had "never had the power and time to write the answer I wanted, namely a proper description of the scientific conscience (bad or good?) of a Quanticist".[60]

In January 1926 the quantum theory had a very different look. As Bohr wrote in a letter to Oseen,[61] it had "passed through a serious crisis in these last years, and at times we have been thoroughly bewildered", but now one could see "brighter days ahead". The new quantum mechanics of Heisenberg, "shaped into such a wonderful theory by Born", and the electron spin hypothesis of Goudsmit and Uhlenbeck had changed the atmosphere completely. "Altogether, the development in this field gives ample material to reflect on the fate to which human conceptions of nature are subjected. We are gradually progressing, I hope, but in every result lurks the temptation to stray from the right path". Bohr was reflecting on how things had gone during the previous decade as he continued:

> "This is so true in atomic theory that at the present stage of development of quantum theory we can hardly say whether it was good or bad luck that the properties of the Kepler motion could be brought into such a simple connection with the hydrogen spectrum, as was believed possible at one time. If this connection had merely had that asymptotic character which one might expect from the correspondence principle, then we should not have been tempted to apply mechanics as crudely as we believed possible for some time. On the other hand, it was just these mechanical considerations that were helpful in building up the analysis of the optical phenomena which gradually led to quantum mechanics".

Altogether Bohr was in an optimistic mood, and glad to share his thoughts with Oseen.

Bohr did not know it, of course, but Oseen had played a major role in the rather involved maneuvering within the Royal Swedish Academy of Sciences that had led to the award of the Nobel prizes in physics for 1921 and 1922 to Einstein and Bohr respectively.[62] But Bohr certainly did

59. See *Works*, Vol. 2, 374.
60. N. Bohr to C.G. Darwin, 14 February 1922, *Works*, Vol. 5, 316.
61. N. Bohr to C.W. Oseen, 29 January 1926, *Works*, Vol. 5, 238-240.
62. Robert Marc Friedman, "Nobel Physics Prize in Perspective", *Nature* 292 (1981) 793-798.

welcome his occasional contacts with Oseen: "I cannot say how great a joy it is for me every time we meet to feel how close we are to each other in our general view of the fundamental problems of physics".[63]

If I had to characterize Bohr's attitude to the state of the quantum theory during its most difficult years, those years when even Pauli wrote that he wished he had become "a movie comedian or something like that and had never heard anything about physics",[64] I would quote a brief phrase from one of his manuscripts. It is not dated but has been assigned to 1924 or 1925 on the basis of its contents, which concern some problematic aspects of the theory of spectra. The fragment ends with the striking phrase: "Much obscurity: great hope".[65]

I am reminded of a famous passage in one of John Keats's letters,[66] where he writes: "Several things dove-tailed in my mind, and at once it struck me what quality went to form a Man of Achievement, especially in Literature ... — I mean Negative Capability, that is, when a man is capable of being in uncertainties, mysteries, doubts, without any irritable reaching after fact and reason", capable of "remaining content with half-knowledge". Bohr possessed this negative capability to an extraordinary degree.

In a letter he wrote near the end of his life, Einstein referred once more to Bohr's approach to physics, capturing its unique quality in a sentence: "He utters his opinions like one perpetually groping and never like one who believes to be in the possession of definite truth".[67]

Yale University

63. N. Bohr to C.W. Oseen, 30 August 1922, in: J. Rud Nielsen ed., *N. Bohr, Collected Works*, Vol. 4, *The Periodic System (1920-1923)* (Amsterdam and New York, 1977) 731. Oseen's very positive response to Bohr's ideas over the years should be remembered when one reads Pauli's characteristically devastating criticism of him and his influence on Swedish physics: W. Pauli to O. Klein, 12 December 1930, in: Karl von Meyenn ed., *Wolfgang Pauli, Wissenschaftlicher Briefwechsel*, Vol. 2, (Berlin, Heidelberg, New York, Tokyo, 1985) 43.

64. W. Pauli to Ralph Kronig, 21 May 1925, in: A. Hermann, K. v. Meyenn, V.F. Weisskopf eds., *W. Pauli, Wissenschaftlicher Briefwechsel*, Vol. 1, (New York, Heidelberg, Berlin, 1979) 214-216.

65. N. Bohr, *Works*, Vol. 5, 171.

66. John Keats to George and Thomas Keats, 21 December 1817, in: Lionel Trilling ed., *The Selected Letters of John Keats* (New York, 1951) 92.

67. A. Einstein to B. Becker, 20 March 1954. Quoted in A. Pais, *'Subtle is the Lord...': The Science and the Life of Albert Einstein* (Oxford and New York, 1982) 417.

COMMENTARY

Dennis G. Dieks *On the paper of Martin J. Klein*

Professor Klein has offered us an interesting look at some of Bohr's considerations and at Bohr's general attitude during the years when the quantum theory was being developed. The point that emerges as particularly striking from his account is Bohr's permanent awareness of the provisionality of his results. More general, Bohr is depicted as an extremely cautious person who found it very difficult to be content with any statement of his point of view; he always felt that the formulation could be improved, and that important points had not received sufficient attention. Professor Klein quotes a draft of a letter to C.G. Darwin that was never sent. As is well known, there are in fact many such examples. Typical is a letter from Bohr to his good friend Oseen of December 20, 1915, which begins with apologies for not having answered a letter of more than a year ago. Bohr writes[1]: "I meant to answer immediately, but I put it off, because I would so much like to explain my point of view in more detail, and I did not think that I was sufficiently clear in my mind about it". Also from other sources and many anecdotes we know how extremely scrupulous Bohr was, particularly in the use of words. As his son Hans Bohr remembered[2]: "Father always pointed out how decisively the human being's place in life depended on the use of words".

However, we must not forget that when it comes to scientific style, boldness, as well as caution, are to be expected with *all* successful researchers. It is no exaggeration to say that the combination of boldness and caution is a necessary prerequisite for any progress in science. As Planck wrote, in his proposal to elect Albert Einstein as a member of the Berlin Academy[3]: "without taking risks it is not possible, even in the most exact science, to introduce real innovations".

1. N. Bohr to C.W. Oseen, 20 December 1915; in: U. Hoyer ed., *N. Bohr, Collected Works*, vol. 2 (Amsterdam and New York, 1981) 564-568.

2. S. Rozental ed., *Niels Bohr. His Life and Work as Seen by his Friends and Colleagues* (Amsterdam, 1968/2) 327.

3. Christa Kirsten and Hans-Günther Körber eds., *Physiker über Physiker*, vol. 1 (Berlin, 1975) 202.

On the other hand, real innovations cannot be mere speculations. So, in the same proposal just mentioned Planck praises in one breath Einstein's formulation as well as his *criticism* of new hypotheses. Passages where such criticism, carefulness and caution are extolled are common in the election proposals of the Berlin academy. Thus Planck speaks of the cautious and conscientious work of Van der Waals[4], and the research of Lorentz is described as carried out with exemplary carefulness and self-criticism.[5]

Without boldness and caution progress is impossible; so it would seem that Bohr in this respect can only differ from his fellow-scientists as a matter of degree. In addition, it might seem that caution and boldness are qualities that first of all express themselves in the personal approach to science and which leave only few traces in published work. They would appear to be attitudes that come to the fore in letters, personal conversations, and so on, more than in scientific results. Professor Klein's account, with its emphasis on the picture which emerges from Bohr's correspondence, might reinforce that impression.

I think it therefore not superfluous to stress that Bohr's "rare blend of boldness and careful consideration", which so impressed Einstein, did not only consist in two traits of character which Bohr possessed to a larger extent than other scientists. In Bohr boldness and caution *together* in a very particular union, constituted a quite unique and distinctive characteristic, not only determining Bohr's personal approach to physics, but also clearly recognizable in his published articles. What Bohr did was to develop a *methodology* in which the conflict of new hypotheses with the older ideas was emphasized (so that the boldness of the new ideas became apparent) whereas at the same time analogies and links with the old conceptions were exploited to the largest extent possible. Caution, in combination with boldness, was so turned into a positive method and did not just result in uncertainty about fundamentals.

To illustrate this point let me attempt to briefly compare and contrast Bohr's and Einstein's approaches to the problems in quantum theory. We have heard a summary of the lecture Bohr held before the Physical Society in Copenhagen in 1913. In his famous report on his discussions with Einstein about epistemological problems in atomic physics, written thirty-five years later, Bohr still referred to this lecture as characteristic for the years when no general self-consistent quantum theory was yet in sight. What is especially striking in the lecture is the combination of two features. On the one hand, Bohr does not get tired of stressing the

4. *Ibid.*, 152.
5. *Ibid.*, 166.

inadequacy of the ordinary mechanics and electrodynamics to account for the experimental findings. Although Bohr emphasizes the coherency of the group of ideas on which these classical theories are based, he is therefore forced to reject them. On the other hand, Bohr assumes that the frequency of revolution of the electron in the stationary states of the atom can be calculated just from those classical formulas. As he puts it[6] "it is quite natural to make this assumption; since, in trying to form a reasonable conception of the stationary states, there is, for the present at least, no other means available besides the ordinary mechanics". We see here Bohr's particular methodology in operation. The classical theories are unsound; this can be seen from very general arguments. In accordance with this the hypothesis of stationary states is introduced, a bold assumption which is in direct conflict with classical theory. However, the elaboration of the new hypothesis proceeds along the lines of the very same theories that are rejected. Of course, Bohr is acutely aware that the basis of the resulting theoretical framework is inconsistent; this is the reason why he continually calls attention to the coherence of the classical theories, which are, however, unfortunately empirically inadequate. This incoherence and the resulting uncertainty about the validity of the scheme are at least partially remedied by a consideration of the limit of large quantum numbers. Here the validity of the results of classical dynamics can be assumed, as shown by the adequacy of the classical radiation law for small frequencies. The connection which is thus established between the new hybrid framework and the classical conceptions makes it possible to make headway in finding relations that must replace the classical ones *also outside* the region of large quantum numbers. What I want to stress is the *particular form* of the *combination* of boldness and caution that thus emerges from a study of Bohr's work. Boldness and caution go together in framing hypotheses which are in patent conflict with classical theory and yet achieve a maximum of correspondence with the classical ideas. This leads to a method in which it is possible to make progress in a guided way.

Let me now briefly discuss Einstein's approach to the quantum problems, which also shows boldness and caution as prominent elements, but in a rather different way. In a series of papers, from 1905 on, Einstein had made it clear that the experimental findings, in particular those represented by Planck's radiation law, necessitated a departure from classical theory which was much more fundamental than most of his contemporaries were prepared to accept.

6. N. Bohr, *Works*, vol. 2, 294.

Einstein's strategy typically involved two basic ingredients: the radiation law, as corroborated by experiment, and very general considerations based on the second law of thermodynamics and its statistical interpretation. The results that are deduced in this way can hardly be doubted, since the whole approach does not depend on any specific theoretical assumption. Having thus demonstrated the inadequacy of the classical mechanics and electrodynamics, Einstein does not pursue the relation between the old theory and the yet to be found new theory, but tries to erect a new, consistent, theoretical framework on the basis of general principles. The closest to something like the correspondence principle he comes in his famous 1916 article on the quantum theory of radiation.[7] There he says that he introduces a few hypotheses which are obtained by carrying over the known situation for a Planck resonator in classical theory to the as yet unknown one in quantum theory. Then there follow the well-known assumptions concerning the probabilities for emission and absorption of radiation by an atom. Typically, the assumptions are chosen in such a way that they involve a minimum of commitment to specific theoretical ideas. The purpose of the whole exercise is to show with such minimal theoretical apparatus that radiation has a definitely non-classical character; it is demonstrated to have particle-like properties — "radiation particles" carrying definite amounts of energy and momentum. As Einstein concludes, this makes the formulation of a proper quantum theory almost unavoidable. But when he attempts to develop such a theory — two years later in an article "Do gravitational fields play an essential part in the structure of the elementary particles of matter?"[8] — he starts from a very general conjecture about a possible connection between the structure of the quanta and cosmology. This line of approach towards the quantum problems Einstein would maintain for the rest of his life. He really puts into practice what he declared in his inaugural speech to the Berlin academy in 1914[9]: the method of the theoretician involves using as

7. A. Einstein, "Zur Quantentheorie der Strahlung", *Mitteilungen der Physikalischen Gesellschaft Zürich* 18 (1916) 47-62.Also in: *Physikalische Zeitschrift* 18 (1917) 121. English translation in: B.L. van der Waerden ed., *Sources of Quantum Mechanics* (Amsterdam, 1967) 63.

8. A. Einstein, "Spielen Gravitationsfelder im Aufbau der materiellen Elementarteilchen eine wesentliche Rolle?", *Sitzungsberichte der Königlich Preussischen Akademie der Wissenschaften zu Berlin* 1919 (Pt. 1) 349-356, on p. 349. English translation in: A. Einstein, *The principle of relativity*, transl. W. Perret and G.B. Jeffery (New York, 1952) 189.

9. Christa Kirsten and Hans-Günther Körber eds., *Physiker über Physiker*, vol. 2 (Berlin, 1979) 245.

a basis general assumptions, so-called principles, from which he can deduce consequences.

Einstein is not lacking in boldness, nor in caution, but in his approach to theoretical physics these elements play a role which is different from the role they play in Bohr's work. Einstein analyses the consequences of the radiation law with great caution, and introduces bold new hypotheses in the attempt of finding a new, coherent, theory. But he resists progressing on incoherent foundations which consist of a mixture of old and new theories.

The distinctive feature of Bohr's work is just the admission of such basically inconsistent combinations of old and new ideas, which are handled in such a way that direct inconsistencies are avoided.

When Pauli wrote about the time he was an assistant to Bohr in Copenhagen he first of all remarked[10]: "Here I had the opportunity to become acquainted with the special scientific methods of this famous researcher". Pauli, of whom it is said that he could not tolerate vagueness or half-truth[11], became a convinced adherent to Bohr's approach. In his well-known review article on the old quantum-theory he reports that Bohr came to see the principle of correspondence not so much as a formal link between the results of quantum theory on the one hand and those of classical theory on the other, but as a purely quantum mechanical assertion which gives a connection between the properties of the light emitted by an atom and the motion in the atom's stationary states.[12] This testifies to what has been said before, namely that classical and new ideas for Bohr came together to become organic parts of one and the same theory, instead of exclusively belonging to two distinct theories that give the same results in some limit situation.

Bohr's endeavour to build a theory on apparently incoherent foundations, while attempting to avoid inconsistencies, eventually led to his doctrine of complementarity. It is sometimes maintained that Bohr formulated his ideas on complementarity as a response, influenced by positivistic philosophy, to the developments in 1925 and 1926. The foregoing brief remarks suggest that the doctrine of complementarity is better seen as the final fruit of a positive methodology which was employed by Bohr from his first research in atomic theory. This methodology is characterized by a quite particular combination of

10. R. Kronig and V.F. Weisskopf eds., *Wolfgang Pauli, Collected Scientific Papers* (New York, 1964) vi.

11. *Ibid.*, viii.

12. *Ibid.*, 315.

boldness and caution, meant as an instrument to make progress *in spite of* half-knowledge, rather than as an expedient to make it legitimate to remain content with half-knowledge.

University of Utrecht

Donald MacKenzie & Graham Spinardi

POLITICS AND TECHNOLOGY IN THE ARMS RACE: A CASE-STUDY IN THE HISTORY OF NUCLEAR WEAPONS SYSTEMS

Introduction

A "new direction in the history of science" — and the history of technology — that is urgently required is work on the influence of military involvement in science and technology. We are all aware of the sheer size of that involvement. In Britain, for example, over half government funding of research and development is military.[1] The historical dimensions of military influence on technology have also been explored to some extent.[2] But on many basic questions we are still shamefully ignorant. For example, the influence of the massive military support of post-war science (e.g physics) on the *content* of the scientific knowledge produced is a question that has scarcely been asked, let alone answered.

When we move to nuclear weapons questions in particular, it is only fair that the public should demand answers from those of us who work in the history of science and technology. Questions like "what shapes science and technology?" and "how can technological developments be promoted or stopped?" arise naturally in this area. True, there has been a great deal of public debate over whether an existing technology (such as cruise missiles) should or should not be accepted. But by the time a technology is ready for deployment a great many issues are settled — fixed in metal, one might say. Debate should surely begin earlier, in the period when technologies are more fluid.

If the history and sociology of science and technology have something to offer the public discussion of nuclear weapons issues, it is therefore perhaps answers to such questions as "what is shaping nuclear weapons

1. See P. Gummett, "What price military research?", *New Scientist* (19 June 1986) 60-63.
2. M. Roe Smith ed., *Military Enterprise and Technological Change* (Cambridge, Mass. 1985).

technology?", "what scope is there for intervention in the processes?", and "does technology push politics, or politics technology?". The answers may turn out not to be simple ones, but as we shall argue at the end of this paper, complexity can, paradoxically, be a source of hope.

'Technology-out-of-control' or 'politics-in-command'?

The understanding of technological change in weaponry is often characterised by two extreme positions. At one pole it is held that technology is out-of-control, and that technological developments drive strategy. Technology is seen as having its own 'internal' autonomous logic which is inexorably driving the arms race. Of course, the more thoughtful advocates of such technological determinism do not impute an entirely autonomous ability of development to technical artefacts.[3] Instead two models have been put forward to explain on-going technological development. One should be called 'technologists-out-of-control' and suggests that technologists and their institutions build up a personal and institutional momentum doing what they enjoy, are good at, get paid for and have invested in. This is what Tom Hughes has called 'technological momentum':

> "loosely connected, mutually reinforcing components that constitute a system of vested interests involving people, institutions, ideas, and artefacts ... that tends to resist change and softly determine the course of events".[4]

'Technology creep', on the other hand, describes a more general effect, whereby the advance of science and technology on a broad front will quietly but inexorably change the strategic landscape.[5] Thus Dietrich Schroeer has argued that progress in computer capabilities "may be a major driving force producing a technological imperative towards improved missile accuracy".[6]

At the other extreme is the view of 'politics-in-command', which is generally proclaimed in official circles in the West, but is also reflected in some antinuclear writing (albeit with a quite different evaluation of the

3. See L. Winner, *Autonomous Technology: Technics-out-of-control as a Theme in Political Thought* (Cambridge, Mass., 1977) for a stimulating general discussion of theories of 'technological politics'.

4. Cited in A. Roland, "Technology and War: A Bibliographic Essay" in M. Roe Smith ed., *Military Enterprise*, 378, n. 220.

5. D. Shapley, "Technology Creep and the Arms Race ICBM problem a Sleeper", *Science* 201 (2 September 1978) 1102-05.

6. D. Schroeer, "Quantifying Technological Imperatives in the Arms Race", in D. Carlton and C. Schaerf eds., *Reassessing Arms Control* (London, 1984) 60-71, on p. 61.

nature of the politics that are in command).[7] Technology is claimed to be a strictly 'dependent factor' in the arms race, and technologists are controlled by political leaders. In this view, elite groups consciously shape technology to meet their goals.

Missile accuracy and nuclear strategy

In this paper, we will test these two competing views by considering a case study of technological change that has great significance for the nuclear arms race — the growth of missile accuracy. Missile accuracy is vital to the operational implementation of nuclear strategy. If nuclear strategy is simply based on the threat of certain and deadly retaliation against cities — the commonly held public understanding of 'deterrence' — then pinpoint accuracy is relatively unimportant. Missiles aimed at the centres of major cities will inflict similar destruction even if they 'miss' by a kilometre. If, on the other hand, missiles are aimed at the enemy's nuclear forces, and particularly at missiles in underground reinforced concrete silos, then very high accuracy becomes crucial. At the same time such 'counterforce' targeting implies an interest in attacking enemy forces before they are used, and raises the spectre of a disarming 'first strike'.

Because of this critical interaction with strategy, missile accuracy provides a key test of the above two opposing views. Are improvements in missile accuracy the result of conscious political direction, aimed at providing the means to implement counterforce nuclear strategy? Or has, as Fred Halliday claims, "the possibility of greater accuracy in targeting missiles *led to* the shift from the 'countervalue' approach, aiming at cities and economic targets, to one aimed at specific military targets, i.e 'counterforce'"?[8]

The case study

Here we shall concentrate on one specific strand of missile accuracy development, that of US Navy ballistic missiles. The many factors involved in missile development vary from country to country, and between different armed services in one country. Attempting to

7. For example, A. Roberts, "Preparing to Fight a Nuclear War", *Arena* 5 (1981) 45-93.

8. F. Halliday, *The Making of the Second Cold War* (London, 1983) 225, emphasis added.

encompass them all in a single comparative study would require a sizable book to do justice to the topic. The choice of the US is dictated by pragmatic considerations. Since the cancellation of Blue Streak in 1960, Britain has purchased its strategic missile systems (minus warheads, but including guidance and associated navigation systems) from the US.[9] And it would be difficult to obtain sufficient technical detail for the Soviet, Chinese and perhaps French ballistic missile programmes.

US Navy submarine launched ballistic missiles (SLBMs) provide the most dramatic example of the growth in accuracy and counterforce capability. The first US SLBM, the Polaris A1, which became operational in 1960, was an archetypal retaliatory countercity deterrent. Apparently invulnerable once at sea, and capable of carrying out the ultimate threat of inflicting immense destruction against cities, it was believed to lack the accuracy to destroy specially protected 'hard' targets. By contrast, the Trident II (D5) due for development in 1989, will be the most important counterforce weapon in the US arsenal. Whereas the Polaris A1 probably had an accuracy of about 250 m over a range of 200 km, the Trident II is said to be expected to achieve an accuracy of 12 m over 750 km range. Between these two were deployed the Polaris A1 (1962), A3 (1964), Poseidon C3 (1971) and the Trident I C3 (1979).

Although official data on such accuracies are classified, reliable estimates can be found in the open literature, as can a wealth of related information.[10] But for our study we have drawn extensively on interviews with those involved in the US Navy Fleet Ballistic Missile programme, especially with its navigation and guidance aspects. This is greatly facilitated in this case by the continuity of organisations involved. The development of all these missiles has been the responsibility of the Navy organisation which was originally set up to oversee the Polaris programme. Then called the Special Projects Office (SPO), to emphasise the urgency of its task, it has since been renamed the Strategic Systems Program Office (SSPO). Similarly design responsibility has remained with the Instrumentation Laboratory of the Massachusetts Institute of Technology, which since its early 1970s separation from M.I.T. has been the Charles Stark Draper Laboratory, Inc.

The interviews were conducted on a wholly non-classified basis, and no security clearance of any kind was sought. This did not prove a major

9. The one significant British intervention in ballistic missile technology in recent years has been the Chevaline programme of alterations to the Polaris A3 missiles. This is the subject of a current study by Spinardi.

10. See, for example, I.B. Cochran and M. Hoenig, *Nuclear Weapons Databook Vol. I: US Nuclear Forces and Capabilities* (Cambridge, Mass., 1983).

barrier to obtaining a quite detailed historical account of the development of Fleet Ballistic Missile navigation and guidance technology and of the forces that shaped key technical and strategic decisions. Indeed such is the wealth of detail obtained for this study that here we will refer to it only by way of illustration of our main conclusions. The detailed case study which substantiates our arguments can be found elsewhere.[11]

So what follows is organised theoretically. Rather than describe the history of the Fleet Ballistic Missile programme chronologically — Figure 1 summarises some key data on the sequence of missiles — we will begin by arguing that no 'hard' technological determinism can explain the development of the Fleet Ballistic Missile programme. The extent to which 'softer' forms of technological determinism are valid will then be explored. The following sections then review the role of various kinds of social factors in shaping Fleet Ballistic Missile technology pressure from technologists command by political elites and the bureaucratic politics of interlocked, competing state organisations. Though we hold this last factor to be very important, we go on to criticise crude 'bureaucratic politics' explanations. Our final section discusses the complexity of the pattern we find in this case-study.

US SLBM Characteristics

	A1	A2	A3	C3	C4	D5
Year	1960	1962	1964	1971	1979	1989
Range(km)	2200	2800	4600	4600[1]	7400[1]	7400[1]
Warhead						
-No.	1	1	3MRV	14MIRV	8MIRV[2]	8MIRV[2,3]
-Yield	600kt	800kt	200kt	40kt	100kt	475kt
CEP(meters)[4]	2500?	2500?	900-1200	450-550	250-450	100-130

1. Off-loading some warheads may allow longer range.
2. A manoeuvrable reentry vehicle intended to help penetration of ABM defences, the Mk-50 Evader has been developed and could be carried by C4 and D5.
3. D5 could also carry up to 14 W76 100kt warheads.
4. CEP, Circular Error Probable, is normally defined as the radius of the circle around the target within which 50% of warheads are expected to fall. It is the conventional measure of missile accuracy.

---

11. Under consideration for publication in *Social Studies of Science*. In what follows, quotations without a citation are from the interview material. More details on who was interviewed, etc., can be found in the full case-study.

Sources

- R. Norris, 'Counterforce at Sea — The Trident II Missile', *Arms Control Today* (September 1985) 9.
- D. Dalgleish and L. Schweikart, *Trident* (Carbondale, Illinois, 1984) 32-33.
- W. Arkin, 'Sleight of hand with Trident II', *Bulletin of the Atomic Scientists* (December 1984) 5-6.
- Interviews.

Technology-out-of-control?

The development of SLBM accuracy and counterforce capability, we found, cannot be explained by simple technological determinism, whereby there is an inexorable, out-of-control, determinate and determining, path of technological development. Not only were technological developments evaluated in terms of organisational goals, but also there were often choices — informed by these goals — between different technical solutions.

Of course, personal preferences would mean that some people — particularly engineers — would like to push technologies as far as they could. With regard to improving the accuracy of Poseidon by using a star-sensor, one Special Projects engineer's attitude was "I could provide it, so I should provide it." But in fact this stellar inertial system was not incorporated into Poseidon, though it is used in the Trident C4 and D5 missiles. Indeed Special Projects Director, Vice Admiral Levering Smith built up a considerable reputation for not choosing technologies simply because they were 'sweet': "Smith's strength was to know if you could do it, it didn't mean you ought to".

Too much weight cannot, of course, be put on this matter of attitudes. Even project managers wholly infatuated with technology cannot do everything that seems technically possible, or even everything that seems technically 'sweet'. Typically, there are choices to be made between different technical pathways. These were much in evidence in this case study. Thus there was a choice between staying with pure inertial guidance or moving to stellar inertial by incorporating a star-sighting. Here the wider political context played a role in augmenting technical doubts which delayed the introduction of stellar inertial guidance till the Trident I. In another case technological developments (especially in computer capabilities) could have led to a switch from a stable platform guidance system to a strapdown one. Stable platforms hold the accelerometers which provide guidance information in a known orientation by means of feedback from gyroscopes. This is the paradigmatic design used in all US strategic ballistic missiles so far. A strapdown system, in which both gyroscopes and accelerometers are simply fixed to the body of the missile, could, on the other hand, be lighter, cheaper to make, and easier to maintain, though computational requirements would be much greater.

Strapdown systems were considered for the Navy missiles, but although computational capabilities were thought to be adequate by the late 1960s, there was doubt over whether strapdown could provide accuracy as good as stable platform guidance, and that possible technical pathway was rejected. Similarly there were choices over the type of accelerometer used — the larger, more expensive, and more accurate PIGAs (Pendulous Integrating Gyro Accelerometers) or the more economical PIPAs (Pulsed Integrating Pendulous Accelerometers) — and the size and numbers of warheads carried.

The technical choices made in the Special Projects Office involved explicit trade-offs between performance and cost, as well as debate about what were the correct goals and what were the best ways to achieve them. For example, the choice of warhead size and configuration in the Poseidon missile was heavily influenced by both strategic and political factors, and was directly related to system accuracy. There was debate over whether the main role of the Navy missiles should be to target urban-industrial areas or hardened military targets, such as missile silos and command posts. Both roles were compatible with the new multiple warhead (or MIRV for multiple independently-targetable reentry vehicle) technology, which Poseidon was the first Navy missile to incorporate. Clearly more than one warhead per missile improved one's force ratios if targeting enemy military targets, and probably meant that a given set of such targets could be covered more cheaply. Indeed that was one of the main rationales which inspired early work on MIRV technology. But MIRVs were also quickly perceived to be useful for assuring countercity retaliation even in the face of anti-ballistic missile (ABM) defences. By multiplying the number of warheads MIRV made it more likely that some could get through to their targets. Broadly speaking therefore MIRVs were compatible with both counterforce and countercity targeting. But the former was generally felt to require high-yield warheads. Optimising hard-target kill thus necessitated large warheads and the best available guidance, both of which were likely to be bigger and heavier. However, with a limited missile size this would mean fewer warheads, or reduced range. Fewer warheads could mean less capacity to penetrate ABM defences, whereas reduced range would potentially increase submarine vulnerability to anti-submarine warfare (because of the reduced patrol area) — in each case possibly compromising the assured destruction/ counter city role.

Within the Navy there was one grouping centred around the Office of the Chief of Naval Operations, and particularly the 'Great Circle Group' set up in 1964 to study strategic issues, which backed a hard-target, large warhead option. In particular they favoured a large warhead known as the 'Mk-17 heavy', which had been proposed by the Air Force to enhance the counterforce capability of their Minuteman ICBMs.[12] Special Projects, on the other hand, preferred the much smaller 40 kiloton Mk-3. Being smaller, more could be carried, so enhancing ABM penetration. In addition Special Projects disliked the idea of a mixed force of Mk-3s and Mk-17s :

> "The Mk-17 was going to be expensive, it was going to require a logistical nightmare .. a specially configured missile assigned to special targets as opposed to submarines which could go on patrol with the flexibility to be targeted from one kind of a target to another without having to worry about what you had loaded in the tubes."

In the end Special Projects got their way, not surprisingly as they were the technical office directly responsible for Poseidon, whereas the 'Great Circle' planners had only a limited capacity to influence the programme. Special Projects also managed to resist pressure for higher accuracy from Defense Secretary Robert McNamara and Director of Defense Research and Engineering, Harold Brown. Although willing to make accuracy improvement a goal, which could if necessary be traded off with other system parameters, SPO Director Levering Smith managed to prevent it becoming a requirement, which had to met come what may.[13]

The reasons for Special Project's lack of enthusiasm for an out-and-out drive to improve accuracy seem to have been threefold. Firstly, a conservative approach to technical problems helped gain Special Project's a reputation for delivering what they promised. Secondly, not pursuing high accuracy and therefore hard-target kill capability avoided direct competition with the Air Force, whose Minuteman force dominated that mission, and 'differentiated' the Navy programme. The Great Circle Group, on the other hand, rather than wishing to avoid competition with the Air Force, wanted to confront them head-on:

12. T. Greenwood, *Making the MIRV: A Study of Defense Decision Making* (Cambridge, Mass., 1975) 55.

13. *Ibid.*, 56. H. Sapolsky, *The Polaris System Development: Bureaucratic and Programmatic Success in Governement* (Cambridge, Mass., 1972) 221-22.

"There were advocates in the Office of the Chief of Naval Operations ... who (felt that) anything the Air Force could do the Navy also needed to do."

Thirdly, the programme was based around the concept of deterrence by the threat of assured destruction retaliation, and many Special Project's personnel became personally commited to this view. As one Officer recalled:

"I'd say most of us saw the role of Poseidon as not different from the role of its predecessors, namely providing an absolutely dependable, reliable deterrent, and most of us were sceptical about the need to dig out hard targets as an essential element of deterrence. We went along with it to the degree necessary in order to keep the program. The nature of democracy is that you're constantly making compromises with different constituencies, and we had to serve the reigning constituency even if we, sometimes felt they were a little nutty."

Technical choices were thus influenced by the 'macropolitics' of US defence policy, by the organisational politics of the Navy and its relationship with the Air Force, also by the 'micropolitics' of the technical community. Whether a star-sensor was needed to supplement pure inertial guidance was doubted by dominant opinion at the Instrumentation Laboratory, where there was a strong commitment to the achievement of ultimate accuracy by refinement of unsupplemented inertial sensors. The stellar-inertial option was pushed by outside industry — the Kearfott Division of Singer.

So there are many instances of technical choices made in the Fleet Ballistic Missile programme, and good evidence of a range of political factors shaping these choices. But 'the social' does not simply operate at the level of preferences between pre-defined technical options. It also shapes the options that are available, and may on occasion actually eliminate the possibility of explicit choice.[14] The social can enter into the definition of what is possible.

One example of this concerns homing re-entry vehicles, a possibility ruled out in the late 1970s in part because adequate testing of the technology was deemed infeasible.[15] Currently guidance only operates

14. See T.J. Pinch and W.E. Bijker, "The Social Construction of Facts and Artefacts: or how the Sociology of Science and the Sociology of Technology might benefit each other", *Social Studies of Science* 1 (1984) 399-441; S. Russell, "The Social Construction of Artefacts: A Response to Pinch and Bijker", *Social Studies of Science* 16 (1986) 331-46; T. Pinch and W. Bijker, "Science, Relativism and the New Sociology of Technology: Reply to Russell", *Ibid.*, 347-60.

15. See the testimony of Rear Admiral Robert H. Wertheim, then Director of SSPO, to the Senate Armed Services Committee, Department of Defense Authorisation Hearings, Fiscal Year 1979, part 9 — Research and Development (Washington, D.C. 1978) 6684.

throughout the first few minutes of missile flight and once released the re-entry vehicles simply free-fall to their assigned targets. Atmospheric perturbations on re-entry set a practical limit to the accuracy thus achievable, and many hold that this limit can only be overcome by manoeuvring re-entry vehicles. The most probable type would home onto their targets by some kind of terrain recognition system, and so would require testing by firing at a large area of land. In practice this would most likely have to be the continental USA and would be politically and legally difficult. To date US strategic ballistic missiles have been tested over water, usually impacting in the Pacific, where indigenous Islanders have so far been unable to bring significant legal or political pressure to bear. So the political power of citizens of the mainland US by comparison with Pacific Islanders is a cause of adequately-tested homing re-entry vehicles being seen as infeasible.

'Soft' determinism – enablement and constraint

Clearly, then, a 'hard' form of technological determinism — in which artefacts follow inevitably from their predecessors — cannot be sustained. At the same time technology is not simply a dependent variable either, as a simplistic 'social shaping of technology' view might have it.[16] Instead technology can sometimes be important as an enabling capability or a limiting constraint.

Technology as enablement amounts to a very 'soft' form of determinism, if it can be so termed at all. Basic technologies and sciences may feed back into specific developments, both allowing and perhaps stimulating technological advances. Thus work in computing, inertial instruments, and geophysics and geodesy has enabled missile accuracy to be greatly improved. Dietrich Schroeer's claim that computing capabilities have driven missile accuracy might seem a case in point. However, as already noted, increased computer capabilities could just as reasonably have led to strapdown guidance systems (had cost and ease of maintenance been more important than accuracy) rather than better performance of stable platform designs. Enabling technologies merely provide possibilities, they need not determine the course actually followed. Rather than Schroeer's 'technological' imperative a more suitable term might be Shapley's 'technology creep'.

16. L. Winner, "Do artefacts have politics?", *Daedalus,* 109 (1980) 121-136, reprinted in D. MacKenzie and J. Wajcman eds., *The Social Shaping of Technology* (Milton Keynes, 1985) 26-38.

Furthermore, advances in some basic technologies and sciences, such as inertial instruments and geophysics and geodesy, have to a large extent been driven by military requirements for more accurate guidance systems. Even where technological and scientific advances on a broader front feed back into specific military technology, as computers seem to have in this case, it is important not to forget how those advances come about, and that they are themselves socially shaped. Widely applicable developments need not be unstoppable if no-one has a particular interest in pursuing them because, paradoxically, everyone may leave it to some one else to pick up the costs. Thus the development of the US Global Positioning System of navigation satellites may have been slowed because all of the armed forces expected to get access to it, but none especially wanted to pay for it.

The development of specific enabling technologies may also be slowed, or stopped, if social conditions are not right. The very high quality inertial components used in strategic missile guidance have few other perceived uses. By the 1960s the best inertial sensors were seen to be already good enough in terms of accuracy for all other applications, such as navigation of civil and military aircraft. Those users instead sought other kinds of technical change, such as lower real costs of lifetime ownership, rather than increased accuracy. More recently even advocates of counterforce accuracy in strategic missiles have begun to doubt whether further improvements in inertial sensors would provide much gain in overall system accuracy. Funds for this form of development of the enabling technology have thus been proving hard to find, and the Draper Laboratory's efforts to develop 'fourth generation' inertial instruments significantly more accurate than the 'third' generation instruments of MX and D5 have not found sources of external support.

Of course, technology may not only provide an enabling capability, it may also prove a limiting constraint. Most obviously, in a complex weapons system such as the Navy ballistic missiles, design interfaces need to be fixed and adhered to. From that point on subsystem technology is constrained by the need to 'fit'. Design decisions may thus have an enduring physical legacy restricting future technology. For example, the physical size of submarine missile tubes (which is of course related to the size of the submarine) limits the size of missile possible. All the US Polaris class submarines built in the 1960s had the same size missile tubes. Once some initial extra space had been made use of, when Poseidon missiles were installed, there remained no space for further expansion. When the Trident I missiles were designed to fit the same submarines, but also to provide about a third more range, this proved a severe constraint, which required much ingenuity to overcome.[17]

But such technical constraints can also be social. When the Trident missile decisions were made, there was considerable debate, over a number of years, about the size of missiles and therefore submarines needed.[18] Admiral Rickover, Director of the Naval Reactors Branch, argued strongly in favour of a large submarine which would require a new nuclear propulsion system. Special Projects, on the other hand, were clearly sceptical of the idea of putting more larger missiles into a bigger submarine. A particular concern was that this means less targets for Soviet anti-submarine warfare, but the question of size was also connected to disputes over the division of responsibility between Special Projects and Admiral Rickover. As it turned out, a larger class of submarine was chosen, Special Projects lost much of its control over the submarine development, and the first submarine, the *Ohio*, incurred considerable time and cost overruns. If instead the need for a larger missile had been refuted and missile tube and submarine size had not increased, then Special Projects may have been able to maintain its control, and minimise Rickover's.

Technologists-out-of-control?

Technological development thus has the potential to follow a number of different courses, rather than one single predetermined pathway. It provides capability and sets some constraints, but is profoundly shaped by social factors. Most immediately, of course, it is shaped by the technologists who develop it, and who have a considerable interest in promoting it. What evidence is there that technologists are, as Lord Zuckerman has suggested,[19] manipulating the strategic and political environment so as to create a market for their favoured technology. Certainly in the Navy ballistic missile programme crucial technological advances were developed externally and before any 'requirement' had been formulated for them. The initial breakthrough of small high-yield

17. See L. Smith, R.H. Wertheim and R.A. Duffy, "Innovative Engineering in the Trident Missile Development", *The Bridge (National Academy of Science)* 1 (1980) 10-19.

18. J. Steinbruner and B. Carter, "Organisational and Political Dimensions of the Strategic Posture. The Problems of Reform", *Daedalus* 104 (1975) 131-54.

19. See, for example, Lord Zuckerman, "Science Advisers and Scientific Advisers", *Proceedings of the American Philosophical Society* 124 (1980) 241-255.

warheads, the introduction of stellar inertial guidance, and the gravity gradiometer installed in Trident submarines are all such cases. However, none found automatic acceptance without careful scrutiny as to how they would affect the goals and interests of the programme. Only if technologists could transform these goals and interests so that they then fitted their prefered technology would the technologists-out-of-control thesis have significance.

Undoubtedly they have tried. For example, one leading guidance technologist clearly perceived that he was trying to engineer people's attitudes as well as guidance systems:

> "I had always been interested in people that I had to deal with and their mental attitudes and why they had these mental attitudes and choosing something that the people in charge had to have."

Not 'wanted', but 'had to have'. There are indeed several instances in the history of the Navy ballistic missile programme where technologists persuaded 'the people in charge' that they had to have something they had not originally wanted — stellar-inertial guidance is one good example. But these are cases where the technologists convinced 'the people in charge' that their existing goals could be better fulfilled by a new device. A full 'technologists-out-of-control' interpretation would require that technologists had gone further than this, and had also changed the goals of 'the people in charge' in attempting to further a prefered technology.

As far as we can tell, this has not happened. One instance of an attempt to do it is in a 1959 paper from leading guidance technologist Charles Stark Draper to the Polaris Steering Group. Draper argued that:

> "fleet ballistic missiles offer many well-known advantages, but will surely be handicapped in competition for national support unless they can be fired with accuracy levels comparable to those of land-based missiles".[20]

But subsequent development suggests that quite the opposite policy was adopted, with a deliberate intent not to compete too directly with the Air Force. When accuracy comparable to ICBMs was eventually sought, in the Trident D5, this came about largely due to other factors.

However, this does demonstrate that goals and interests are not fixed. Moreover, technological possibilities clearly can alter the way that they are perceived. Improvements in guidance accuracy did not drive the Navy to adopt a counterforce philosophy, but they made accomodation to such

20. C.S. Draper, "Submarine Inertial Navigation — A Review and some Predictions", paper presented to the Polaris Steering Group on 2 October 1959 (Draper Laboratory Library, CSD-107) 3.

a policy much easier. As technological possibilities evolve so goals and interests need constantly to be reevaluated. Occasionally radical innovations in technology may lead to equally radical reappraisals of goals and interests, as occurred with the very origins of the Navy ballistic missile programme. During the 1940s interservice disputes the Navy and the Air Force had supported exactly opposite targeting strategies:

> "In urging the procurement of the new long-range B-36 bomber, which would necessarily have limited bombing accuracy, the Air Force was advocating a policy of using nuclear weapons against large urban areas in time of war. The Navy challenged both the effectiveness and morality of this strategy and offered as an alternative increased reliance on carrier-based aircraft to pinpoint and deliver nuclear and conventional weapons to military targets."[21]

But by the mid 1950s the Air Force had started to use counterforce roles as a means of justifying huge bomber forces, and the Navy had begun to advocate a policy guided by "an objective of generous adequacy for deterrence alone (i.e., for an ability to destroy major urban areas), not by the false goal of adequacy for 'winning'."[22] In a word — Polaris! Technology had clearly played a major role in shaping social interests, as well as *vice versa*.

In general, though, technology and technologists have not otherwise played a major role in shaping the goals and interests of the Navy ballistic missile programme. The momentum the programme has acquired — in, for example, the inexorable process of 'follow-on' from the A1 to the A2 to the A3 to the C3 to the C4 to the D5 — is an institutional and political momentum as much as residing in technology in the narrow sense. The perceived interests of the programme as a whole have dominated, and the enthusiasts of particular technologies have not been able to refashion those interests at will.

Politics-in-command?

What then of the view that technical change is simply directed by politics-in-command. On the face of it, there is evidence to support this view. For example, pressure from the Office of the Secretary of Defense — especially Secretaries McNamara and Schlesinger and Director of Defense Research and Engineering Harold Brown — was important in the improvement of missile accuracy. Most obviously Secretary Schlesinger's 1974 review of nuclear strategy towards more flexible Limited

21. Sapolsky, *Polaris System Development*, 5-6.
22. In D.A. Rosenberg, "The Origins of Overkill Nuclear Weapons and American Strategy, 1945-1960", *International Security* (1983) 3-71, on p. 56-57.

Nuclear Options was especially significant.[23] Planning to use selective nuclear strikes in order to make deterrence more credible, to gain more political leverage, and leave the possibility of stopping short of an all-out exchange created a 'need' for more accurate weapons.[24] According to one SSPO Officer, Schlesinger "just kept pushing for improved accuracy." Unwilling to commit themselves to a stringent accuracy requirement, SSPO's leadership agreed to implement a programme geared towards achieving a better understanding of error sources in Navy ballistic missiles so that, at some future date, significant accuracy improvement would be possible. Mainly based on better testing instrumentation, this was known as the Improved Accuracy Programm, and ran from 1975 to 1982, at a total cost of about $600 million. It was this programme which provided the knowledge which was claimed to make Trident II accuracy comparable to land-based ICBMs.

However, it is clear from this example that this was not a question of command in the archetypal military sense. 'Pressure' is indeed the appropriate term here as political leaders sought to obtain greater accuracy from the Navy missiles. For two reasons it was simply not possible to command the technology into being.

First of all there is a general limitation to command. Technological change simply does not possess either the transparency or the predictability that would be required for straightforward command to be possible. Dominant social groups, whether these be political, business or military elites, are typically in no position to shape technological change in the conscious, literal sense that we can imagine an artist moulding clay. If they choose between given technological options, then those who present those options to them have an opportunity both to set the agenda (deciding which options to present) and to influence the decision in the way they portray the advantages and disadvantages of different options. If they seek to create a new technology, then they may well need advice as to feasibility — for what they desire may be physically impossible, or hopelessly expensive, or whatever. Sometimes, indeed, an elite group may simply be unsure precisely what its interests and goals are, at least in the area in question.[25]

Secondly, and more specifically, politics-in-command fails to provide

23. D. Ball, *Dejà Vu: The Return to Counterforce in the Nixon Administration* (California, Seminar on Arms Control and Foreign Policy, December 1974).

24. J.S. Wit, "American SLBM Counterforce Options and Strategic Implications", *Survival* 24 (1982) 163-74.

25. This is one finding of work on the participant observation of technological change being conducted at the University of Edinburgh by Anne Elder.

an adequate explanation in the US political context. Here a combination of democratic political system and complex and overlapping jurisdictions creates a situation where political leaders such as a President or Secretary of Defense have to engage in a process more akin to bargaining than to giving orders.

Successful technological programmes in the US political system are often characterised by astute manoeuvring. Technological success depends on programmatic success, and programmatic success depends on maximising support and minimising interference. The importance of the relationship between various organisations within and outside the US government, and of the constant jurisdictional bargaining, has been well captured by a body of work on 'bureaucratic politics'.[26]

Bureaucratic Politics?

Studies of 'bureaucratic politics' emphasise that what happens is the outcome of a process of struggle between different parts of the 'state'. These struggles can be conceived as occurring at various levels, though in reality such a distinction may not be so clear cut.

Inter-service rivalry between the Navy and the Air Force is the most obvious level of struggle, but interestingly has been conspicuous by its absence during much of the Navy ballistic missile development. Many in the Navy sought to avoid competition with the Air Force by adopting a policy of 'differentiation' which gave the Navy ballistic missile force the distinct role of Assured Destruction retaliation. During the 1950s national nuclear strategy was the subject of heated (and sometimes public) debate between the armed services. In 1959 the Air Force put forward proposals for coordinating all US strategic nuclear weapons under one command, which naturally enough would be them. As it turned out the Navy was able to retain its command over Polaris, but the new warplan (known as the SIOP or Single Integrated Operational Plan) of 1960 effectively put into practice the Air Force doctrine and institutionalised Strategic Air Command dominance over the targeting process.[27] Following this setback, and in view of the Navy's other

26. In addition to Sapolsky, *Polaris System Development* and Greenwood, *Making the Mirv*, M.H. Armacost, *The Politics of Weapons Innovation the Thor-Jupiter Controversy* (New York, 1959) and G.T. Allison, *Essence of Decision: Explaining the Cuban Missile Crisis* (Boston, 1971) provide other classic perspectives.

27. See F. Kaplan, *The Wizards of Armageddon* (New York, 1983) 263-270 and Rosenberg, "Origins of Overkill", 50-69.

commitments, there was little to be gained by direct competition with the Air Force. Instead the main focus of Navy policy was to protect the FBM programme.

But, of course, not everyone in the Navy wanted to have Polaris in the first place, if it meant that they would have to make do with a smaller surface fleet. Of all the US armed forces, the Navy probably is the most traditional in outlook and in the 1950s both submarines and strategic weaponry were still peripheral to that tradition. Intra-service bureaucratic politics thus played a part. Indeed resistance within the Navy to the idea of a Navy ballistic missile probably hindered the initial acceptance of the idea in the mid 1950s, just as Air Force attitudes slowed the development of the ICBM.[28] The early success of SPO owed much to the patronage of Chief of Naval Operations Admiral Arleigh Burke and to the skilful bureaucratic manoeuvring of its first Director Admiral 'Red' Raborn. Burke gave Polaris the highest priority and told Raborn that "if more money is needed, we will get it."[29] As persuasively described by Harvey Sapolsky, Raborn and SPO then set out to create conditions in which their hand-picked team could rapidly develop Polaris.[30] Sapolsky notes four strategies which contributed to SPO's bureaucratic, and consequently programmatic and technological, success: as already noted, differentiation of a special role, which was represented as being of crucial national importance; co-option of potential critics and disruptive elements; moderation of short-term goals in order to maximise long-term support; and managerial innovation to provide an aura of efficiency, however spurious. In the post-Sputnik atmosphere of the late 1950s and early 1960s skill at bureaucratic maneuvring and technical ingenuity combined to forge a successful new technology - Polaris.

More recently, however, the SSPO has not been able to maintain its independence from the rest of the Navy. Whereas pressures for direct competition with the Air Force could be largely resisted in the mid 1960s they carried more weight a decade later. Other parts of the Navy clearly were more interested in a Navy counterforce missile able to compete with hard-target kill capabilities of the Air Force. By 1983 the careful differentiation of Navy missiles from those of the Air Force had ended.

28. See E. Beard, *Developing the ICBM: A Study in Bureaucratic Politics* (New York, 1976) and R.L. Perry, *The Ballistic Missile Decisions* (Santa Monica, California, October 1967; RAND Corporation P-3686).

29. Chief of Naval Operations Admiral Arleigh Burke to Rear Admiral Clark (Op-51) and Rear Admiral Raborn (Office of the Secretary of the Navy), 2 December 1955, reproduced in Lockheed Missiles and Space Company, Inc., *The Fleet Ballistic Missile System* (Sunnyvale, California, n.d.) 6-7.

30. Sapolsky, *Polaris System Development*, 41-60.

Not only was Trident II scheduled to have roughly comparable hard-target kill capability, but also Chief of Naval Operations, Admiral Watkins, felt happy to compare them directly:

> "By 1991 we believe you could have 4 to 5 D5 equipped Trident submarines, which is more than the equivalent of an MX field in terms of hard target kill capability."[31]

Finally, bureaucratic disputes could also be important in questions of the internal organistion of SPO (SSPO) itself. Building such a complex weapons system involved dividing it up into various subsystems, and carefully defining the boundaries and interfaces. For example, Special Projects branch SP-23 was responsible for guidance and fire control, whilst SP-24 was the technical branch for navigation. Since an inertially guided missile must know its whereabouts at launch, navigation information is crucial for the functioning of the guidance/fire control subsystem. One issue was whether the guidance/fire control subsystem should have to request data from the navigation computer, despite the possibility of delays, as SP-24 wanted, or whether navigation should 'broadcast' positional data several times a second to the rest of the system, which was SP-23's preference.

Thus at various different levels 'bureaucratic politics' has played a role in shaping technological change in the Navy ballistic missile programme. Shrewd management in the early days ensured that SPO capitalised on the political and strategic environment to establish a nearly autonomous organisation within the Navy, so facilitating the development of Polaris. More recently the SSPO has been unable to maintain such independence, and suffered considerable loss of control in the arguments over the Trident submarines, which led to the setting up of a separate Trident Office within the Navy. This not only affected the technology which was produced, but may also have contributed to the problems that were encountered producing it.

Interestingly one group of potential actors in the 'bureaucratic politics' arena, the large aerospace corporations, do not in this case seem to have exerted undue influence, as suggested by the more conspiratorial versions of Military-Industrial Complex theory. The fact that, once initiated, the existence of the Navy ballistic missile programme has never been seriously in doubt may have limited the incentives for dubious or even corrupt behaviour. The prime contractor for all the US fleet ballistic missiles has been Lockheed. Though other parts of Lockheed have sought contracts by questionable means, and the resulting scandals have

31. W.M. Arkin, "Sleight of Hand with Trident II", *Bulletin of Atomic Scientists* 40 (December 1984) 5.

affected the West German, Japanese and Italian governments as well as the Netherlands royal family, no such pressure has, to our knowledge, been brought to bear in the case of the fleet ballistic missile programme.[32]

Problems with 'bureaucratic politics'

The 'bureaucratic politics' model captures much of the complexity of organisational interaction which result in the formulation of policy, and the production of technology. But it is not entirely satisfactory, at least in its cruder expressions. In particular there are four major factors which need to be woven into a 'bureaucratic politics' framework to provide a satisfactory explanation of technological change in US Navy ballistic missiles.

First of all, neither the Executive Branch nor Congress have been irrelevant, as a simplistic reading of 'bureaucratic politics' might imply.[33] Thus not only were various leaders in the Department of Defense influential in pushing for improved accuracy, but also some members of Congress, most notably Senator Brooke of Massachussets, were able to effect some restraint on it. His particular efforts in the late 1960s seem to have played a role in deferring the introduction of stellar inertial guidance from Poseidon to Trident I. In another instance President Nixon directly intervened to speed up Trident development in October 1971, apparently in an attempt to strengthen his bargaining position and to bolster domestic support in the run-up to the first SALT agreement.[34]

Second, 'bureaucratic politics' is sometimes taken to mean that only domestic considerations matter in weapons policy. 'The enemy', 'the Russians', plays a role only as an image conjured up by domestic forces — and conjured up in the form that best suits those domestic forces. This is undoubtedly partially true, but taken to the extreme of totally discounting all 'external' factors it is neither theoretically nor empirically supportable. Theoretically international relations and intelligence concerning Soviet behaviour are just another level at which factors can feed into a 'bureaucratic politics' model. What is to distinguish the 'external' factors of alliance politics (within NATO, for example) from 'internal' machinations within the USA. Empirically, it seems beyond dispute that

32. See A. Sampson, *The Arms Bazaar* (London, 1977), Chapters 6 and 13.

33. For a sophisticated presentation of 'bureaucratic politics', see G.T. Allison and F.A. Morris, "Armaments and Arms Control Exploring the Determinants of Military Weapons", *Daedalus* 104 (1975) 99-129.

34. Steinbruner and Carter, "Organisational Dimensions", 140.

Soviet behaviour has made a difference to the US Navy ballistic missile programme. For example, estimates of current and future Soviet ASW and ABM capabilities have played an important part in design decisions. Of course, both these estimates and the 'correct' responses to them are shaped by the social context — the restricted range of the UK Polaris A3 has not caused the British government undue concern over the vulnerability of a much smaller submarine fleet to Soviet ASW, some fifteen years after the decision in the USA to move to an extended range missile was justified by reference to this 'threat'. Indeed, because of the lack of experience, the hypothetical nature of nuclear war planning, and the uncertainty of much intelligence data, many of the 'facts' concerning nuclear weapons policy are extremely negotiable, and policy implications even more so. In practice, however, there is closure around many facts and Soviet behaviour is not irrelevant. At the most basic level, had the Soviet Union not built ICBMs, and then put them into hardened silos, the US controversy over hard-target kill capability could hardly have taken place.

Thirdly, any 'bureaucratic politics' model must take into account the wider political context, as well as looking at the organisations directly involved in weapons development. During the 1970s and 1980s attitudes towards nuclear war and arms control drifted away from the liberal 'dovish' stance — typically supportive of assured destruction versions of nuclear strategy coupled with arms control — towards more conservative 'hawkish' views which favour counterforce nuclear strategies coupled with defensive systems, and are extremely sceptical of the value of arms control agreements. Usually adherents of assured destruction place the main emphasis on preventing nuclear war by the threat of massive retaliation, and (often implicitly) see little other political utility for nuclear weapons. On the other hand, supporters of counterforce targeting and Star Wars type defences consider it important to be able to threaten credibly to fight and (in some sense) win a nuclear war because of the perceived political leverage this would engender.[35] Curiously the Navy ballistic missile programme has had the 'best of both worlds' in this shifting situation. The traditional combination of invulnerability and poor accuracy made SLMBs the ideal deterrent weapon in many liberals' eyes, and ensured support for Polaris and Poseidon. Many 'doves' did question the need for Trident when the concept was first proposed — mainly on the grounds that another 'follow-on' system was premature

35. The most articulate proponent of this view is Colin Gray. See, for example, his "Nuclear Strategy: the Case for a Theory of Victory", *International Security* 4, no 1 (1979) 54-87.

and simply unnecessary.[36] However, although the delays and cost overruns of the Trident submarine programme attracted criticism, liberal concerns were mainly diverted away from the Trident missile to its Air Force counterpart, the MX ICBM. Even Trident's counterforce capabilities can, in this situation, appeal to Congressional 'doves', because the argument that "Trident can do anything that MX can do" can then be deployed against MX.

'Contextual influence' should not, however, be taken as a given. It is variable, and importantly related to the chances of success of a given weapons programme. The contrast between Trident and MX is instructive here. The problems which befell MX are largely due to the efforts of its advocates during the 1970s which pushed the issue of the MX basing mode into prominance. By forcefully drawing attention to the increasing vulnerability of the Minuteman force to Soviet ICBMs, and by allowing the MX to be linked to an alternative basing rationale, they raised the 'technical' issue of basing mode to an unusual level of political prominance. Congress and the Executive played active roles in determining not just the strategic ends, but also the technical means. Whereas Trident D5 remained a technological 'black box', unopened and unquestioned at the technical level, the MX basing mode issue became 'political'.

This made the MX an easier target for opponents of counterforce, so deflecting opposition away from the D5, and led to the great delays in the MX programme. Just as the great autonomy of Special Projects had enabled Polaris to be rapidly developed, so the lack of autonomy of the MX programme made it vulnerable to delays and interference. Avoiding such interference and keeping the 'black box' unopened requires 'bureaucratic politics' which are responsive to the wider political context.

Fourth, and finally, 'bureaucratic politics' as ordinarily understood tends to underestimate the significance of the production of military technology. Organisations like the Special Projects Office do not simply have to cope with an obdurate 'social' world (as 'bureaucratic politics' might be taken to imply), they also have to cope with an obdurate 'material' world.

The significance of the process of production, and its socioeconomic nature, are now well understood for civilian technologies[37], and should not be overlooked in the military sphere. Guidance system components are amongst the most demanding of human products, and for the fleet

36. See D.D. Dalgleish and L. Schweikart, *Trident* (Carbondale, 1984) 57-59.

37. See, for example, the studies reprinted in part two of MacKenzie and Wajcman, *Social Shaping*.

ballistic missile programme they have to be made in considerable (if not huge) numbers to extraordinarily high standards of reliability.

Issues of producibility and reliability have undoubtedly shaped the way technical choices were made in the Navy ballistic missile programme. One example, is the choice of electrostatic gyro (ESG) which was developed to provide better navigation for the Trident submarines. Honeywell's design — which incorporated a large hollow electrostatically supported metal ball — was believed by some to provide the theoretically best potential accuracy. However, the system choosen was a small, solid-ball design produced by the Autonetics division of Rockwell International. Not only was the Honeywell design inherently more difficult to produce, but the company compounded this difficulty by shifting their production facility at a crucial time. As one SSPO engineer recalled:

> "We told the corporation what was going to happen to them — that they were going to move down there, half of their people weren't going to go, and this [gyro production] was an artistic thing... They could produce them, but it wasn't something you could put on a production line. You had people who had techniques, etc... They moved down to Florida and nothing worked. Half the people didn't move, wouldn't move, some of them retired, everything we said happened in spades. Eventually they were in deep [trouble]... [Autonetics] persuaded us that since Honeywell was falling on their face, we ought to give them [Autonetics] a chance, and we decided that we would. And they funded much of that chance themselves, and the answer is that we had a time when we wanted to see the Honeywell thing on the [U.S.S.] Compass Island [the ship used for testing navigational equipment], and when that time came Rockwell had one there and knew how to run their thing and ... Honeywell delivered one a few months late and they hadn't the slightest idea what to do with it..."

Thus even though system accuracy was at an increasing premium, and the large-ball ESG believed by many to be potentially more accurate, the more producible small-ball design triumphed, aided by the producer of the large ball forgetting the socially delicate nature of human skill.

Complexity and hope

The case study surveyed here, the development of guidance and navigation in US Navy ballistic missiles, provides no simple picture, and allows no easy theoretical explanation. Technical change appears to be the outcome of a complex, interactive, and sometimes contradictory process. There simply is no single dominant determining factor which can be distinguished. Indeed the social and the technical are hard to differentiate, and paths of influence are neither unidirectional nor stable in their effects through time.

This complexity may be the theorist's despair — at least until such time as a framework of thought that makes clear sense of it can be found. But we would also suggest that it should be the citizen's hope. The complexity of the world of nuclear weaponry is, perhaps, simply the complexity of all the products of human activity. The very fact of complexity ought then to be a reminder that it is human activity — not a gigantic impersonal determinism — that has given rise to the nuclear world. As a human creation, that world is beyond neither our reason nor our influence.

Furthermore, because the world of nuclear weaponry is the outcome of several levels of often contradictory social and political processes, it is potentially more susceptible to change than it might appear. A relatively small political force (such as Senator Brooke's campaign against counterforce) can, if applied at the right place and time, achieve results that are far from negligible. A better understanding of the complex social dynamics of nuclear weaponry can, we would hope, increase the chances of such events happening more often and having a more permanent effect.

University of Edinburgh

COMMENTARY

John Grin *On the paper of Donald MacKenzie and Graham Spinardi**

Introduction: the need for a theoretical framework

Understanding armaments dynamics is important for gaining more effective political control over it, which in its turn is a necessary, although not sufficient, condition for stemming the arms race. Therefore the lack of an interactive and more comprehensive model that already in the seventies was considered to be necessary to replace earlier monocausal explanations[1] is an important problem.

MacKenzie and Spinardi fully recognize the shortcomings of monocausal explanations. Nevertheless, they attempt to tackle the general problem of understanding armaments dynamics by comparing the material from their interesting and well-documented case-study, which covers all relevant factors and actors, with these limited explanations. From the viewpoint of getting the more general model they agree to be needed, this means a very heavy reliance on induction. A more fruitful approach might have been to construct the rough contours of an interactive and more comprehensive model, and subsequently use the case-study to test and eventually to refine it.

An obvious objection against such an approach is that it seems virtually impossible to construct a model of any general value that both is 'broad' enough to take into account the inherently stochastic nature of the process of technological development and all the factors influencing it, and gives some useful insight into this process. This problem may be circumvened, however, by defining 'useful insight' in a way much less ambitious than 'predictability of innovations': the model, to be useful, should in concrete cases offer policy makers an insight into the various possible paths of technological development and the impact each of these

*The author wishes to thank Dr Marlies ter Borg and Dr Herman de Lange for their useful comments.

1. See e.g. J. Yanarella, "The "technological imperatives" and the strategic arms race", *Peace and Change* 3 (1974) 3-16.

would eventually have. With this it will make a more effective political control possible in the sense that choices can be made, so that technological developments will not dictate policy.

The above-mentioned criterion still has another advantage. It links up well with an alternative way of formulating the main problem, one that seems more sensible than asking whether "politicians can shape technological change like an artist can mold clay": can technological developments be politically controlled (in the sense indicated above) or not? This formulation has the merit of focussing on the point in which the essence of the whole class of 'politics-in-command' theories differs from the essence of the whole class of 'technological imperative' models.

From the above and from studies like that of De Lange[2] it is clear that the model required will be something like a general framework, which is broad enough to include at least all the factors covered by earlier monocausal explanations and is to be filled in a different way for each case. Fortunately, recent literature[3] seems to offer opportunities for such a general model of technological change. Below, an attempt is made to adapt this model to the area of military technology, and to compare it with the material from the case-study of MacKenzie and Spinardi. Finally, the question whether this rough attempt meets the above mentioned criterion of utility is dealt with.

The rough contours of a general model: an attempt

The general model referred to above consists of two components. The first is what will be called here the *technological paradigm*. It can be viewed as a set of positive and negative heuristics, and determines the paths of technological development. Within the paradigm, some paths are more probable than others; these paths will be called *technological trajectories*. These technological trajectories can be quite general (e.g. mechanization and automatization) or specific to the particular sector involved. They have a certain momentum of their own, and can give rise

2. H. de Lange, *De bewapeningswedloop tussen de Verenigde Staten en de Sovjetunie 1945-1983. Een onderzoek naar de relatie tussen politieke en militaire macht* (Leeuwarden, 1983).

3. R.R. Nelson and S.G. Winter : "In search of a useful theory of innovation", *Research Policy* 6 (1977) 36-76; G. Dosi, "Technological paradigms and technological trajectories", *Research Policy* 11 (1982) 147-162.

The possibility to use these concepts for explaining armaments dynamics was suggested by Hans Akkermans in: H. Akkermans et al., *De Patriot – schot in de roos of dure misser*? (Amsterdam, 1985) 81-90.

(on their own, or combined with other trajectories) to new applications which can eventually lead to a new technological drive.

Thus the first component of the model concerns the generation of new technologies over time. The second one involves the selection of those ideas that will evolve into successful innovations: the selection environment which determines how relative use of different technologies changes over time. The factors, making up this selection environment can also — in a more long-term oriented way — influence the technological paradigm.

A comparison with biology may be helpful. Within a population, occupying a certain ecosystem, genetic mutations are continuously being generated. The ecological environment influences the result in two ways: first, by selecting the direction of mutation, and, second, by selecting among mutations in a more darwinian manner. In an analogous way, the social circumstances both influence the long-term selection of the technological paradigm and the selection of the development and application of technologies within it.[4] And of course, new selected innovations, like mutations in biology, on their turn influence the environment.

Below, an attempt will be made to draw the rough contours of a specialization of this general model to the specific area of military technology. In advance, it should be noted that in this area non-market features are more relevant that in most other technological areas.

The military area, the technological paradigm is characterized, amongst others, by typical features like 'survivability', 'reliability under adverse conditions', 'interoperability' and so on.

As regards the selection environment, a first element is formed by the goals of the 'producers'. These are not only defined by monetary profit, but also in terms of technological challenge. Second, the reaction of the military, voters (including the peace movement) and legislators is relevant. Here available budgets are of course an important factor. Furthermore, factors like the prevailing military concept[5] and the dominant image of war[6] play a role, as do factors such as arms control policy (be it by forbidding certain directions of R&D and stimulating alternative ones, or by creating a need for 'bargaining chips'), perceived adversary copabilities and interservice rivalry. Of course, these factors may interact.

4. Dosi, "Technological paradigms"(note 3), 156.

5. Depending on the particular area of R&D involved, this concept can be the strategic or the operational doctrine.

6. F.O. Miksche, *Vom Kriegsbild* (Stuttgart Degerlöch, 1976).

A final element of the selection environment is formed by the innovation diffusion mechanisms. These can be investment by the investing institution and imitation by others. In a non-market context they depend on the degree of information and value (non-)sharing among institutions.

Comparison with the fleet ballistic missile guidance case

Within the scope of these comments, it is of course not possible to compare this roughly sketched model extensively with the material from the case-study by MacKenzie and Spinardi. In this section it will be attempted, however, to make plausible that the model is able to explain at least part of the developments in the fleet ballistic missile case.

The existence of a technological paradigm within the instrumentation lab is suggested by phrases like "technologies that *participants* perceived as radically different from those employed on Polaris began to be advocated by 'outsiders' to the set of *interconnected organizations* that worked on Polaris" (italics by JG). Technical caution was part of this paradigm because of the participants' interest not to spoil their reputation, as becomes clear from the section on the Poseidon guidance development. Survivability was another element of the paradigm, as is reflected by two important technological trajectories: increasing range and increasing the number of (independently targetable) re-entry vehicles. With this, two trajectories have been mentioned. A third one is increasing accuracy while decreasing the size of the guidance system.

This paradigm was shaped by the elements of the selection environment. The requirement of survivability for instance was inspired by the dominant image of war which, especially for the Navy, was modelled to a large extent on the Pearl Harbour experience.[7]

The operation of the selection environment on choices within the paradigm can also be illustrated in the fleet ballistic missile guidance case. Increasing accuracy of the guidance system, a technological trajectory from the fifties up to the present, was for a long time exploited with nuclear retaliation as the prevailing military concept. For instance, in spite of its higher accuracy PIGA was rejected as the main gyroscope for the Polaris A3, both for higher costs and worse producibility, and for changing Soviet capabilities. An increased anti-submarine warfare capability made longer range necessary, and this required a larger propulsion system which implied that the guidance system had to be

7. Fred Kaplan, *The Wizards of Armageddon* (New York, 1984) 92-110.

smaller. A similar mechanism played a role again with the Poseidon, in which also the last PIGA was replaced by a PIPA, because the increased (perceived) anti-ballistic missile threat made Multiple Independently-targetable Re-entry Vehicles desirable, whereas PIGA did not make sense in these MIRV's. Changes in the selection environment led to changing priorities. Later on, in the Trident D5, the PIPA's on their turn were replaced by PIGA's, not only because the latter had, in the meantime, become smaller, but also because accuracy was by then seen as an important feature for the military concept that had become more war-fighting oriented.

Also the history of the star-tracker is an interesting example of how the selection environment operates. The star-tracker was pushed by Singer for market reasons. This innovation encountered a diffusion barrier. Because the system was at that time, at least psychologically, linked with a hard target capability that did not fit in the military concept of retaliation, it was finally rejected as part of the guidance system of Poseidon. Later on, it was accepted because in combination with the increased range of Trident C4 it led only to a minor increase in overall accuracy, while on the other hand the longer range (required because of improved Soviet anti-submarine capabilities) could, with the star-tracker, be realized without less overall accuracy. Finally, in Trident 5 the star-tracker contributed to the overall accuracy which was desired in the new concept of war-fighting.

So, with retaliation as the prevailing military concept, more accurate systems were either rejected for the benefit of other systems that scored better on other points, or combined with other, neutralizing trajectories that were encouraged because of, for instance, perceived changes in the adversary's capabilities. Only when war fighting became the military concept this trajectory was accepted as a tool for improving total system accuracy, leading to a hard target kill capability.

Thus one may conclude that the theoretical concepts discussed in section 2 are capable of explaining some of the developments described in the paper of MacKenzie and Spinardi. Therefore it may be worthwhile to perform a more fine-mazed analysis. But even if further refinements would lead to a model that can satisfactorily explain this and other historical cases, an important question remains: is the model useful for gaining a more effective political control?

A future-oriented perspective?

As was suggested in the introduction, to be useful, the model should be able to give insight into the (near) future options of technological developments. Therefore, it would be necessary, first, to identify technological trajectories within the prevailing technological paradigm. For this the contemporary history of technology offers good opportunities. Second, the trends in the military capabilities they present should be summed up. Third, the selection environment should be mapped out, making use of the findings of the social history of technology. Here it should be noted that some of the elements of the selection environment should be treated as variables; for instance, the technological options should be identified both for the dominant military concept and for concepts developed by critics thereof.[8]

So the social history of technology can contribute both to the refining of a model of technological development, and to the application of the model in contemporary cases; thus supporting attempts to gain more control over military technology.

All this may not be a 'new trend' in the history of technology. But it certainly can be an interesting, and, from the viewpoint of stemming the arms race, a relevant one.

Free University, Amsterdam

8. See e.g. Marlies ter Borg & Frank Barnaby eds., *Emerging technologies and military doctrine – a political assessment* (London, 1986) in which new military doctrines are evaluated in two doctrines: planned NATO-doctrine and alternatives developed by critics thereof.

Herbert Mehrtens

MATHEMATICS IN THE THIRD REICH: RESISTANCE, ADAPTATION AND COLLABORATION OF A SCIENTIFIC DISCIPLINE

What I mean with the concept "Third Reich" in the title of this lecture seems to be clear. To be precise I would have to present my conceptions of the power structure of the Nazi state, its ideologies, its historical changes and the place of science and education therein. But for the concept "mathematics" some sociological and historical clarifications are required which are not that easily at hand. How does mathematics relate to politics? Historically the time of National Socialist rule supplies many examples of such relations. Mathematicians invented the idea of a racially specific "German mathematics", they have produced collections of exercises in school mathematics laden with ideology, they did statistical research on the inheritance of schizophrenia and "social unfitness", and they developed theories and numerical methods for the determination of the optimal shape of artillery shells and the V2 rocket. But is all this really "mathematics"? Or, to invert the question, what is left, if all this is not mathematics?

As a historian I follow the smell of humans. I am looking for mathematics where mathematicians are. And mathematicians did what is listed above. To give a more sober and more precise definition, I take mathematics to be a social and a cultural system at the same time. The cultural system is the living and growing body of the specific type of knowledge we call mathematics. It is alive only if there is a social system of communication within which this knowledge is handled. Thus the cognitive and the social side of the discipline are just two sides of the same coin. I call this coin the social system of mathematics. As a social system its primary function is the production of new mathematical knowledge. The social system defines itself through its knowledge. As a system it has a cognitive identity which we call "mathematics" in the pure sense, which appears not to have any relevant relations to practical politics. The social system also needs a further identity, a social identity, since mathematicians at work are also, e.g., teaching, giving lectures, editing journals — activities that are not mathematics in the pure sense. To be a mathematician one has to produce new mathematical knowledge, all other activities might be necessary to earn a living as a mathematician but they are no necessary conditions in the professional self-definition.

Much could be said about the structure of knowledge producing social systems, their cognitive and social identities, their relation to their environments, the necessity to take the individual mathematician as part of the environment of the system, and the interactions of the system with its political environment. But I shall try to exemplify the concepts by applying them in a historical outline, which will look first at the relations of mathematics to Nazi-ideology and then follow the history of the institutions of mathematics as related to their political environment. As to the latter topic just one more general remark is needed. Sciences like mathematics have no immediate market for their products. Therefore they have to organize themselves in a way that the necessary material and cognitive resources are ensured. In twentieth-century Germany this means that mathematicians are with rare exceptions employed by universities and institutes of technology (*Technische Hochschulen*). There they have their mathematical institutes, which are part of the social system of mathematics as well as part of the social system of academic education. The fact that here two social systems interpenetrate is important because the educational system in Germany is administered by the state. In contrast, disciplinary institutions like scientific societies or journals have a much greater autonomy. The dependence of mathematics on society at large for its resources and in our case especially on agencies of the state enforces specific forms of political adaptation. The sciences need public legitimacy and they need the support by those systems which consume their products like the neighbouring disciplines, technology, industry or bureaucracy. So much for generalities. Let us begin with the story.

In 1936 Ludwig Bieberbach, professor of mathematics at Berlin university, 50 years old and well established in his field, started a new journal called *Deutsche Mathematik* which was to "serve and foster the German character in mathematics". In a series of lectures and papers Bieberbach explained in some detail what he meant by the "German character" in mathematics. He developed a theory of psychological types of mathematical work which was basically dualistic. On the one hand, there was the good "German" type of doing mathematics. Such a "German" mathematician had an intuitive substantive understanding of what he was talking about, related it to the real world, and, with the spirit of a fighter, struggled with his material to form it into a good and meaningful mathematical theory. On the other hand, there was the "counter-type", tending to lose the connections to the real world, doing mathematics completely for the sake of itself, juggling with abstract, meaningless concepts. Although Bieberbach took some care not to identify the counter-type explicitly with the Jews, the implicit identification was clear enough as soon as he related his psychological types to racial types. His examples of the bad mathematics of the counter-type were Jewish and French mathematicians. And Bieberbach's followers were indeed very explicit in that respect.

Bieberbach's key concept was "intuition", a vague enough term, which centered around the use of geometrical imagery. The opposite term was "formalism": abstract, logified, formal mathematics, that presented itself without any reference to the real world or to intuitive imagery. One of his examples was the formalist, and Jewish, mathematician Landau who, in a textbook, had introduced the number pi with a definition by the power series expansion without any reference to the concrete geometrical meaning of pi. The good German author would, naturally, take things the other way around and start from the concrete, keeping the geometrical image along with the concept of pi. Another of Bieberbach's favourite examples is Gauss' introduction of the complex numbers interpreted in the number plane, which Bieberbach called "organic and concrete" and saw as the decisive step to grant the complex numbers "citizenship in mathematics". The "ungerman" contrast to Gauss Bieberbach presented with the "symbolism" of Cauchy's algebraic treatment of the complex numbers.

Bieberbach did not succeed in convincing his fellow German mathematicians of his image of mathematics. Nor was he able fo find sufficient backing from the political powers to force himself and his views into a dominant position. After a complicated struggle he was forced in 1935 to resign from his position as permanent secretary of the German mathematicians' association, mainly because the ministry preferred not to intervene too strongly in the association.

What were the motives for Bieberbach's ideology of a "German" mathematics? For one thing most of Bieberbach's colleagues seemed to believe that he was "subjectively honest", that he believed what he said and wrote. In fact, when I met him shortly before his death, he still held that there were distinct styles in mathematics connected to psychological types, and in an apologetic letter of 1949 he hinted at his interests before 1933 and added that he had tried to "kindle the interest of those in power" to care for the safety of himself and of his colleagues. Thus there were at least three motives involved. One was probably a feeling of insecurity and fear, maybe also some pressure in his family, which made him adapt to the new order quicker and more willingly than necessary.

The second was the attempt to legitimate mathematics vis-à-vis the new powers by showing that mathematics, despite its intellectual aloofness, was a good German thing to do. Such legitimating arguments are quite explicit in Bieberbach's contemporary writings. He specifically defended mathematics against the utilitarian approach, arguing for pure mathematics as the right stuff for the German character. Further he defended mathematics against the Nazi physicists, especially Philip Lenard and his followers, who despised mathematical physics and mathematics with it, and who appeared to gain considerable influence in the first years of the regime.

These two motives, personal interests and the attempt to legitimate mathematics by adapting it ideologically, were, so to speak, the boundary conditions, since Bieberbach did not just put up a rhetorical facade. He believed most of what he said, and the ideas he had were older. In 1926 he had given a lecture in which he exposed the difference and opposition of two "ideals of mathematics" or "styles" quite in the same vein as in 1933 and after, but without races, psychology and politics. In 1926 Bieberbach opposed the formalist style to what he called the "intuitive concreteness". His formulations then very sharply condemned formalism as the "removal of sense and meaning from mathematics" by a "paradoxical over-estimation of logical form". Formalism, he argued, would be disastrous to the application and to the teaching of mathematics. Intuitionism on the other hand was to him "the escape from a nightmare". Intuitionism, Bieberbach wrote, "attempts to account for the fact that human beings do mathematics, and it attempts to allow for intuitive concreteness everywhere in its constructions".

Bieberbach addresses and attacks a triple phenomenon which he does not keep apart. He talks about the two opposing philosophies of mathematics, formalism and intuitionism, about the styles of mathematical exposition, but above all about the "modern" mathematics of the day, for which the abstract and highly formalized modern algebra was paradigmatic. These three developments were indeed interrelated and can be interpreted as an effect of what might be called the modern revolution of mathematics. Since the last decades of the 19th century mathematics had indeed turned largely to formalism in Bieberbach's sense. Sets of elements became the object of mathematics, which had no other meaning than to carry mathematical structures. The structures themselves were formally fixed in axioms without any reference to a meaning or to applications outside the mathematical theory itself. And the language of mathematics was logified and formalized, in order to be completely precise. There was indeed no more room for imagery and ambiguity, at least in the official presentation of mathematical results. For the theoretical sophistication of mathematics, its precision, and its possibilities of further development this was a great gain. The meaning of mathematics for anything outside itself, however, became almost invisible. During the twenties it became quite clear that this was the leading movement in mathematics, carried, especially in Germany, by the most famous and most powerful mathematicians and institutes. If we may call this movement the "modernization" of mathematics, then there was also a wide resentment against mathematical modernism. Bieberbach was by far not the only one to oppose what he and others called the "formalist style", but he was one of few who were explicit about it and the one who took the most spectacular road into Nazi Germany.

In the second half of the twenties Bieberbach was also involved in a struggle against the mathematicians of Göttingen university, the stronghold of mathematical modernism. These struggles centered around questions of general politics, namely the relations to French scientists who had been leading the international boycott of German science after World War I in the field of mathematics. Thus matters of general politics, struggles for professional power in the discipline, and the opposition of two images of mathematics were mingled in those developments. However, Bieberbach lost all his battles. Modern mathematics gained even more influence and weight, formalism became the prevailing foundational program for mathematics, Göttingen mathematicians kept their dominant position in the discipline, and the majority of German mathematicians opted against chauvinism, at least in science.

Bieberbach, like others, was a loser. His turn to Nazism then might be interpreted as a late attempt to an anti-modern revolt in mathematics, a scientific counterrevolution with the help of the political powers. This was possible because the ideology of National Socialism was addressed at such anti-modern sentiments. It promised the loser of modernization in all fields of life the return to a safe and meaningful world. No wonder, that such appeals also worked in mathematics, like it did in physics. Bieberbach and his colleagues could indeed in a meaningful way relate to Nazism. The loss of meaning through modernization, the loss of visible connections between the various parts of the world of life were phenomena that especially Germans after World War I experienced very strongly, mathematicians included. This is part (and only a part) of the explanation of the rise of National Socialism, and it rose everywhere, also in the exact sciences.

Let me add a remark of further interpretation. First, the intrusion of politics in this case came into the social system of mathematics through individual mathematicians, who constructed for themselves an image of their science and the political world opposed to the prevailing positions and structures. Nazi ideology of mathematics was produced by mathematicians themselves, but not by the social system of mathematics. It was rather the need of individuals to find a way to integrate their private, political and professional worlds which led to these absurdities. The motives of the absurdities, however, were not absurd. A modernized science, especially mathematics, puts the burden of the task to find a meaning in the science and the rest of the world onto the individual mathematician. This problem is by no means confined to the sciences. It is one of the basic problems of social and cognitive modernization.

German Fascism gathered the energies for an anti-modern revolt, but it did not offer a solution to modernization problems. To come to power and to remain in power National Socialism had not only to live with the powers of modernization, it had to use them, like in the case of the modern means of propaganda or the war machinery or its collaboration with big industry. The National Socialist revolution never took place as a social revolution. Would-be revolutionaries like Bieberbach in mathematics served to drive conservative forces into cooperation with the apparently more rational parts of the regime, but they never gained any considerable influence themselves.

In applied fields and in matters of education mathematicians were quite willing to serve the regime in order to survive. The rhetorics of utility was easy enough to handle and did not imply substantial changes of research practice. There was hardly anybody who could judge whether a piece of mathematics was relevant for application or not. More alarming is the willingness of mathematicians to offer mathematics as a carrier of ideology in school education. Since universities were the most important base institutions for mathematics, and the main educational task there was the training of secondary school teachers, school mathematics was the single most important field of disciplinary politics since the 19th century. The *Mathematischer Reichsverband*, an association of mathematical societies, was the group that cared especially for these interests. It immediately declared its loyalty to the new regime in 1933 and did everything to keep the number of curricula hours at the given level. Any conceivable type of rhetorical adaptation was used to achieve this aim. But there was more than rhetorics. The association produced a handbook for teachers with ideology-laden exercises of every kind. "Mathematics at the service of national education" was its title. It was a joint enterprise of university professors and secondary school teachers. There is, for example, the following exercise: "To build a lunatic-asylum required 6 Million Reichsmark. How many new homes at cost of 15.000 Reichsmark each could have been built for this sum?" Obviously here a simple calculation exercise carries a hidden message about social values.

But there is more to it. The calculation is not just a carrier of values, it appears as an evaluative tool. Calculating costs is offered as the appropriate approach to questions of social politics.

The handbook offers not only the mathematics lesson as a good place of the not-so-hidden ideological curriculum. Mathematical instrumentality is advertised as a possible element of the National Socialist world-view, and through this mathematics education and the science of mathematics are legitimated. This observation is so alarming because in the destruction of the Jews, in eugenic politics, in general social politics such an approach of an instrumental, calculating rationality played, in fact, a very important role.

It used, however, merely traditional, elementary means. The most sophisticated approach was the attempt to establish a central data-processing institution during the war with a punch-card system with the ultimate aim of numbering every person and every nut and bolt of the *Reich*. Such tendencies hardly influenced the science of mathematics. One of the few exceptions is mathematical statistics which was only in a few cases developed and applied to such ends. One mathematical statistician, for example, published extensive studies into the inheritance of schizophrenia and of "social unfitness". He explicitly offered statistics as a sharper and better tool for social and medical policy than the "vague theorems of biology". In these cases mathematics came much closer to the ideology and practice of National Socialism than in Bieberbach's theories of styles, types and races. The problem in this case is that the social system of mathematics or of any other science gives room for such options. Quality, productivity, and autonomy of mathematical research are hardly endangered by ideological curricula and problematic applications. On the contrary, they serve to ensure the existence and growth of the system in such a political environment. The social system, with its cognitive identity reduced to the core of purely mathematical production does not enforce moral or political awareness in its practitioners. The question of moral evaluation has to be applied to the individual mathematicians and their behaviour. I know of one mathematician who visibly rejected the workings of that association which produced the quoted handbook. Carl Ludwig Siegel's Frankfurt mathematical seminar resigned membership of the association in 1934. In the same year, however, the society for applied mathematics and mechanics joined.

The institutions of mathematics, institutes, societies and journals were under pressure and resisted or adapted to the regime in various ways. I shall survey some of these developments briefly in the following part of the lecture. The strongest pressure on mathematics occurred at the institutes in universities and institutes of technology. The expulsion of

Jews and political opponents started early in 1933 and created an atmosphere of fear and uncertainty. Along with the expulsions came denunciations and threats, a quest for politicizing academia from the student body, part of the staff and from the ministry. Otto Blumenthal of Aachen for example was denounced by the student council as a "friend of communists" and as a Jew. He was taken in what was called "protective custody" for a fortnight in April 1933, released and soon after dismissed from the university. He emigrated to the Netherlands only late in 1939. After the occupation he was deported to Theresienstadt where he died in 1944. These few data refer only to one of the many stories of suffering, declining hope, and despair. Most of the mathematicians who remained in Germany and who were not exposed to persecution appear to have been able to blind themselves, so that they did not see the suffering of their colleagues and friends. This is not specific to mathematics, but neither is mathematics an exception. The universities adapted rather quickly and with little resistance to the pressures. The orders from the government were taken as legitimate and handled bureaucratically and submissively.

To give an idea of the extent of the expulsions: of some 230 teachers of mathematics in academic institutions almost a third had to leave: 26 of the 95 professors and more than half of the lecturers (*Privatdozenten*). There were hardly enough young mathematicians to fill these gaps. Furthermore, the number of students shrank dramatically. The most famous institute of mathematics in Göttingen had 432 students in 1932. In the summer of 1939, still before the war, there were no more than 37. The politics of the ministry of science and education towards the universities led to a loss of positions in mathematics of about 10%. In the course of the employment of new personnel there was a rather strong pressure to employ convinced National Socialists during the very first phase of the regime. But there was also, and constantly growing, in mathematics and the natural sciences at least, the tendency to apply established standards of scholarly quality. In Göttingen for example the first two newly-employed full professors were an excellent and politically loyal mathematician, Helmut Hasse, and a less-than-mediocre mathematician, Tornier, who was an ardent Nazi. The decisive position as the director of the institute, however, was given to Hasse, the outstanding mathematician, and not to the outstanding Nazi. Very quickly conflicts developed, and the ministry attempted to find a position somewhere else for Tornier, but the universities defended themselves with radically negative judgments on Tornier's mathematical qualities. Finally Tornier was transferred to Berlin, which appears like a last resort for outspoken Nazis among mathematicians. By 1938 Göttingen was restaffed fairly well with some high-quality personnel, which were in general politically completely loyal without being radicals. Similar stories may be told for most institutes. The situation by 1938 was such that basic loyalty to the state was granted while professional quality prevailed as the criterion for evaluation of personnel. The general conditions and the lack of young scholars, however, led to a clear decline in quantity and quality of mathematical work.

While the pressure on the state-controlled institutions was strong and led to a compromise between political demands for loyalty and professional standards of quality, the non-state institutions of the discipline were more independent. The expulsion of Jews from societies and editorial boards was officially ordered only in 1939, when there were hardly any Jews left. The publishing house Springer, for example, most important for mathematics, tried to keep its publishing standards by a careful politics of compromises, and had to submit to the pressures to a large extent only from 1938 onwards. Clearly, there was some room to resist quick adaptation.

Of the societies of mathematics the *Reichsverband*, which I mentioned before, was the one that adapted to the principles of the new regime most completely and quickly. It hailed the *Führer* and the new Germany early in 1933 and adapted the principle of leadership (*Führerprinzip*) and a paragraph to its statutes excluding "non-Aryans". The reason for this submissiveness lies not only in the individuals involved. It was rather the nature of the association as an interest group oriented mainly at the link between the science of mathematics and school mathematics. Since the school system was tightly state-controlled and furthermore teachers tended quite strongly towards National Socialism it seemed clear that a policy in the interest of the discipline was only possible from an affirmative position. The handbook I have quoted and other activities of the association show quite clearly that the motivation was not one of general political convictions but rather of lobbying in a salesman's spirit. With the general trend this association shifted by 1936/37 from heavily ideological movements to legitimation of mathematics by utility. It lost in the course of years, however, its importance because on the one hand, the association of teachers of natural science and mathematics was nazified, and on the other hand, the German mathematicians' association took over more and more the interest policies for the discipline.

The mathematicians' association (*DMV*) had kept quiet during the first year. Apparently the submissive moves of the closely related *Reichsverband* presented enough visible adjustment, so that the *DMV* could try to keep its character as an unpolitical scholarly society. There was, however, the permanent secretary and editor of the society's journal,

Ludwig Bieberbach, who made his political orientations only too explicit by publishing an ugly letter in the journal in which he answered a Danish critic of his ideas on styles and races, the mathematician Harald Bohr. This publication did not have the consent of the president of the society nor of Bieberbach's fellow editors. Bieberbach's open letter was clearly a threat to the international reputation of the society. The conflict became open and had to be solved at the society's meeting in the fall of 1934. At this meeting the society furthermore faced the motion to adopt the principle of leadership and make Bieberbach *Führer* of the German mathematicians. The society was unable, at this meeting, to keep the matter one of professional conduct. In the case of Bieberbach's open letter it went as far as to "condemn" Harald Bohr's publication "insofar as it contained an attack on the new German state", while it merely "regretted" Bieberbach's letter and the form of its publication. The motions concerning the leadership principle, however, were rejected, and a clever compromise was found. A clause was adopted which looked like such a principle and gave the new president the right to appoint and dismiss the officers of the society. The intention of this clause was clearly to dismiss Bieberbach from his office. To become valid this clause had, however, to go through a procedure of legal recognition, and Bieberbach as the secretary was responsible for such matters. Furthermore the clause contained some trivial mistakes. All this led to a sharp conflict within the council of the society during the following months. In the end Bieberbach had the ministry intervene, but the result of the intervention was that Bieberbach as well as the president of the society resigned their offices. The society did not have to change its statutes but it offered an informal declaration of loyalty. It looks as if Bieberbach had functioned as the threatening radical driving the prey into the arms of the ministry. After this was achieved Bieberbach was caged in and did not get any share of the prey. Such procedures appear to have been quite frequent in those years, they can be, e.g., detected also in the history of the association of German engineers.

In the following years the *DMV*, collaborating with the *Reichsverband*, was very concerned about the establishment of new university curricula for teachers and for what were called "industrial mathematicians". Clearly this was an attempt to legitimate the discipline by underlining its utility and to find arguments for new positions in the universities. The emphasis was on applicable mathematics and not applied mathematics. As a matter of fact there is no indication that the society intended to expand the field of applied mathematics. Rather it was an adjustment of the teaching task without strong repercussions for the structure of research. Similarly the ministry supported the development of such a new curriculum, and finally in 1942 it was introduced, along with similar curricula in physics, psychology and other disciplines. But there is hardly a sign of an explicit policy by the ministry to build up the applied side of mathematics in research.

One of the innovations in mathematics under the Nazi regime was the "industrial mathematician" and the Diploma in mathematics, qualifying for positions outside the educational system.This was a technocratic modernization, but it was achieved not only by the combined efforts of mathematicians and the ministry but also by the destructive Nazi politics. The extreme shortage of personnel in science and engineering and the exclusion of many people from the civil service in universities and schools for political, racial and other reasons led to the appearance of industrial mathematicians in industry and research institutes. This trend was reinforced by the war, when many mathematicians sought safe positions in institutions where military research and development was done.

The *DMV* sealed its compromise with the state when it decided that its president be reelectable. Thus Wilhelm Süss, not an outstanding mathematician but a clever politician for his discipline, became permanent president of the *DMV* from 1937 to 1945. With this decision the *DMV* had in a sense adopted the *Führerprinzip*, which enabled it to act in a politically effective way. Süss' activities were especially successful during the war. By 1942 the political leaders were willing to take any last straw in reach, and the opportunities of science and engineering grew enormously after they had been neglected almost completely during the first two years of the war. Süss managed to set up a program for mathematical literature, handbooks of applied and applicable mathematics useful to engineers and natural scientists. Very clearly this effort was not directed at putting mathematics immediately at the service of the war aims. None of the books focussed directly on the military tasks. All of them were of a general kind, valuable and even of very high quality in any other situation beyond the war and Nazi-rule. Süss also attempted the foundation of a research institute for mathematics with the examples of the Princeton Institute for Advanced Studies or two recently founded Italian institutes in mind. In the end Süss was able to exploit the competition between three state agencies involved in science policy. In 1944 the ministry of science and education granted permission and funds for the foundation of a *Reich* research institute of mathematics. It did not even bear the name "institute for applied mathematics" which Süss had announced earlier. Indeed, when late in 1944 mathematicians started to work at that institute in Oberwolfach in the Black Forest there was no military research involved. Süss and the *DMV* had during the war gained recognition for mathematics as a pure, applicable science. It appears as if the interaction between scientific politics, the inaptness of political and military leaders, and competition and rivalries between state agencies had made it possible for the war to be put in the service of science, and not vice versa.

During the war the field of applied mathematics was mainly represented by Alwin Walther, head of the institute for practical mathematics at Darmstadt. Walther had built up this institute since the mid-thirties and developed it into a center for practical mathematics in the sense of developing and applying calculating devices for scientific and technological tasks. He became the leading figure in the field and cooperated with Süss in mathematical politics. In Walther's institute a large number of different assignments from the military and industry were handled, among them calculations for the V2 rocket. Like Süss, Walther was searching for long-term developments rather than attempting to put his institute directly at the service of the war. His aims were not those of the war. This is illustrated by a remarkable story.

The SS had under Himmler become more and more involved in the economy and had begun to take over also parts of military research and development, linking it to its system of the concentration camps as a resource of slave labour. The SS proposed a mathematical research institute in 1943, which was to be set up in a concentration camp to make use of the mathematicians and scientists among its inmates. Both Süss and Walther welcomed the plan and promised their cooperation. Süss also attempted to save the life of a mathematician through this cooperation. Walther, as the applied mathematician, promised to train the head of the SS-institute and offered further collaboration by his institute. The institute was established in the concentration camp Oranienburg-Sachsenhausen and began work in the fall of 1944 with 18 inmates of different nationalities. The cooperation with Walther was not very effective because the Darmstadt institute had been destroyed by bombs by this time. Walther however, in planning for the SS-institute, had proposed that it take over tasks of "long-term importance" because, he wrote, everyone else was concerned with matters of greater urgency. I can only comprehend this proposal and the whole collaboration as a pure survival technique for the discipline, exploiting any possibility.

Both Süss and Walther survived as representatives of their disciplines. Both were quite influential in the first post-war years. What they had done was not rated as collaboration with a criminal regime but rather as an effort for the sake of the discipline against the regime. And their scientific and organizational competence was also accepted and utilized by the victorious allies, who were very interested in the results of the German mathematicians, scientists and engineers during the war. The survival strategy of the system of mathematics, namely to show its utility while defending its professional autonomy was effective also in getting over the threshold of 1945. As such mathematics was as relevant or even more so to the victorious powers as to the Nazi regime.

This observation leads back to a last story, namely that of the society for applied mathematics and mechanics (*GAMM*) in the early years of the Third Reich. It conceived of itself as an unpolitical and very useful body of scholars which needed autonomy and international relations to be effective and useful. To ensure this position it used the power of the air ministry, which indeed cared more for planes than for races. The president of the society, Ludwig Prandtl, was quite successful in trying to keep the politics of blood and soil out of his realm. But he was only able to do that because he gained a central position in aircraft research. This, in his definition, was not political but just the honest work of a scientist and engineer. Such a way of being unpolitical and even to some extent resistant against Nazi politics was well accepted by large parts of the regime and by other power blocks in society, like industry and the military. In the end, however, the society for applied mathematics could not escape the developments of the State. The international congress for mechanics did not accept the invitation to Germany extended in 1938, because the society had been unable to obtain the complete disregard for races from the ministeries of education and propaganda. And in 1940 the very last Jewish name was taken from the list of members of the society.

The story illustrating the type of professionalism displayed in this case is the following. In 1933 two members of the council of the society proposed to resign because they were Jewish. Prandtl as the president wrote a letter to Erich Trefftz, an excellent applied mathematician and definitely no friend of Nazism. Prandtl proposed to resign himself too and asked Trefftz to take over the society. Trefftz urged Prandtl to remain in office because otherwise the society would be in danger. And furthermore Trefftz added emphatically: "If we are forced to exclude Jewish members, I would hold the dissolution of the society to be the most honourable reaction". Prandtl answered at length, arguing that the society was necessary to give the field its appropriate place between mathematics and physics. This need, he wrote, "persists today more than ever [...] To my mind this has nothing to do with considerations of honour, since it concerns simply a necessity of the discipline". Prandtl remained in office, and in fact managed to secure quite a share of the new resources provided for military development for mechanics and applied mathematics.

Prandtl argued with the "necessity of the discipline" against an appeal of political morality from Trefftz. Indeed, the discipline, as a social system, is interested in its own survival and in further evolution. It does not care whether its members are Jewish or Aryan, persecuted or honoured, as long as the system is not in danger. And people like Prandtl, Süss and Walther are the appropriate representatives of such a system. Looking at the three societies and at how their leading figures changed or remained one might conjecture that even if the majority of mathematicians were, for moral and political reasons, against the regime, systemic inertia forced them into a collaboration which in effect supported the regime. The basic need for mathematical education and research was present in Nazi Germany despite all regressive ideology of blood and soil. The social system of mathematics related to this element of its environment and ensured its survival in the ways I have discussed. It rejected the type of politicization which would have endangered its productivity, and it struggled to keep its international dimension. Insofar the discipline was resistant. But it did not offer any resistance to the expulsion and the destruction of Jews and others, nor did it strengthen the moral or political opposition of its members. This is talking of the system, not the individual mathematicians who have to be judged individually. There were those who saw what happened and opposed the regime. But they did not hold positions representative for the social system. They were excluded from such roles not by the political rulers but by the system itself. The one positive element about mathematics and natural sciences as social systems in political environments that I can discern in this story is that the international dimension gives a strength to the system to resist national developments endangering the discipline. On the other hand, the systems depend on the national societies, and in Germany obviously state control of the academic institutions enforced political adaptation to a large degree. The cognitive identity of mathematics, its universality and its purity remained largely untouched. This cognitive identity is a very powerful center of the social system, but it is merely an abstract, ideal center. The concrete life of mathematics was profoundly affected and influenced in that period and mathematics was a small but inexpendable part in the functioning of that regime and society for the twelve years of its existence.

Technical University, Berlin

Acknowledgement and Bibliographical Remarks

The research on which this lecture is based has been supported by a grant from the *Stiftung Volkswagenwerk*. A book on the topic is in preparation.

The stories told here, and the quotes given, have been presented in more detail in the following papers:

H. Mehrtens, "Anschauungswelt versus Papierwelt — Zur historischen Interpretation der Grundlagenkrise der Mathematik" in: H. Poser and H.W. Schütt eds., *Ontologie und Wissenschaft, philosophische und wissenschaftshistorische Studien zur Objektkonstitution* (Berlin, 1984) 231-276. (Analysis of the so-called modern foundational crisis of mathematics, related to the prehistory of Bieberbach's "German mathematics").

H. Mehrtens, "Die "Gleichschaltung" der mathematischen Gesellschaften im nationalsozialistischen Deutschland", *Jahrbuch Überblicke Mathematik* (1985) 83-103. English transl.: *The Mathematical Intelligencer*, forthcoming (History of the integration of the three mathematical societies, *Reichsverband*, *DMV*, *GAMM*, into the Third Reich).

H. Mehrtens, "Ludwig Bieberbach and "Deutsche Mathematik" in: E. Phillips ed., *Studies in the History of Mathematics*, (MAA Studies in Mathematics 26, The Mathematical Association of America 1987) 195-241. (Analysis of Ludwig Bieberbach's career and ideological and political commitments).

H. Mehrtens, "Angewandte Mathematik und Anwendungen der Mathematik im nationalsozialistischen Deutschland", *Geschichte und Gesellschaft* 12 (1986) 317-347. (History of applied mathematics in the Third Reich, including the creation of the "industrial mathematician" and the history of the two research institutes of 1944).

H. Mehrtens, "The Social System of Mathematics and National Socialism: A Survey", *Sociological Inquiry* 57 (1987), 159-182. (Sketch of the theory of the social system of a scientific discipline applied to history of mathematics in Nazi Germany).

H. Mehrtens, "Das soziale System der Mathematik und seine politische Umwelt", *Zeitschrift für Didaktik der Mathematik,* forthcoming. (Detailed analysis of the social system of mathematics and its relations to politics).

On the history of mathematics in Nazi-Germany see also:

M. Pinl and L. Furtmüller, "Mathematicians under Hitler", *Year Book Leo Baeck Institute* 18 (1973) 129-182.

S. Segal, "Mathematics and German Politics: The National Socialist Experience", *Historia Mathematica* 13 (1986) 118-135. (Erratum *HM* 14 (1987) 297).

R. Siegmund-Schultze, "Theodor Vahlen — zum Schuldanteil eines deutschen Mathematikers am faschistischen Mißbrauch der Wissenschaft", *NTM Schriftenreihe für die Geschichte der Naturwissenschaften, Technik und Medizin* 21 (1984) No. 1, 17-32.

R. Siegmund-Schultze, "Faschistische Pläne zur "Neuordnung" der europäischen Wissenschaft: Das Beispiel Mathematik", *Ibid.* 23 (1986) No. 2, 1-17.

N. Schappacher, "Das mathematische Institut der Universität Göttingen 1929-1950" in: H. Becker et al. eds., *Die Universität Göttingen unter dem Nationalsozialismus: Das verdrängte Kapitel ihrer 250 jährigen Geschichte* (München, 1987) 345-373.

COMMENTARY

Hans Freudenthal *On the paper of Herbert Mehrtens*

Mehrtens' paper is, indeed, representative for a new trend in the history of mathematics. From the dawn of civilisation onwards mathematics has been socially relevant, more overtly in olden times when mathematics was straightforwardly applied in business and administration, and somewhat later, in astrology, than it is nowadays thanks to a variety of applications through the multifarious mediation by natural and social sciences. It is also a fact that quite early much and much more diversified mathematics was produced than was strictly needed by the society, and that producers neither thought about, nor cared for, the use of their production, nor could they predict future uses. Moreover, for a considerable time, mathematics has been, and even now is, appreciated as a teaching subject because of its alleged selective value for further study and employment.

There exist other social features of mathematics research and education but its social relevance, and one of them, considered by Mehrtens, is the community of mathematicians, particularly its being embedded in society at large. Whereas in other sciences, in particular in medicine, the professional community — surgeons, midwives, dentists — has been a cherished subject since long, the much larger community of mathematicians has not been a subject of study, except perhaps from the viewpoint of a curiosity.

Although the present historical interest in the community of mathematicians has been kindled by more recent historical events, there is still a number of problems and questions left regarding the past that is worth being studied and answered. The social reputation acquired by mathematics in the past century is at least partly rooted in politics. Among the *Grandes Ecoles* created by the French revolution in order to outstrip the *ancien régime* universities, the most brilliant was the *Ecole Polytechnique*, which counted among its professors and graduates the greatest French mathematicians. Never in his life could Hermite, who had finished second in his year's *concours* of admittance, swallow the disappointment about his being rejected because of a limp in his gait. Indeed, the *Ecole Polytechnique* was a kind of military school, and in the 19th century the military superiority of the French was often explained by the mathematical spirit and education of their officers — remember Napoleon's well-known foible for mathematics. Does mathematics owe at least part of the reputation it acquired in the 19th century in continental Europe to this military explanation of its effects? To what degree did it contribute to the flourishing of mathematics at the universities and to the coming into being of High Institutes of Technology?

I must apologise for this digression, which I made only to put Mehrtens' paper into the broader frame, which for that matter he himself had indicated at the beginning of his lecture.

Rightly Mehrtens started his more detailed account with the story of Bieberbach's success and failure. He restricted himself to what in Bieberbach's adventures is relevant to his chosen subject. Let me add a few personal recollections. At ours in Berlin Bieberbach was known as a moderate leftist, at least if compared with the rightist majority in university chairs. His "doctor father" had been the Jew Schoenflies, to whom he was also indebted for his academic career. Once somebody in Berlin told me the only communists he ever knew as such, he had met in Bieberbach's home. Still as late as 1 April 1933, the famous boycott day, thus months after Hitler's take-over, Bieberbach voiced his sympathy for his dear colleague and friend Issai Schur, whose absence he complained in public. I think Mehrtens' explanation of the riddle of Bieberbach's change of mind is correct: First, Bieberbach's psychological instability due to a lack of selfconfidence — it was impossible to face him because if addressed he cast down his eyes. Second, his naive belief that under the given circumstances, mathematics could be saved and served by stressing its Teutonic genuineness. As a mere anecdote I would like to mention that during the famous Bieberbach-Bohr affair, Brouwer spoke about Bieberbach as the blond negro — an exaggerate but somehow fitting characterisation of Bieberbach's physical appearance.

Distinguishing mathematical styles — Bieberbach's point of discussion — is an interesting but thorny subject. Bieberbach's attempts, however, were quite superficial and their combination with race characteristics flatly mistaken and self-contradictory.

Mehrtens rightly signalises a triple confusion in Bieberbach's approach, but unfortunately he adds a fourth source of misunderstanding by sticking to the terms intuitionism and formalism, which properly used apply to foundations of mathematics rather than to mathematics itself. If one agrees that there is a variety of mathematical styles, one has first of all to distinguish between style of research and style of presentation of research results. If one compares modern mathematical papers with those of, say, a century ago, one is struck by the difference between the often informal discursive style and easy readability of old mathematics and the streamlined and heavily formalised style of ours, which to be understood asks as it were for decoding. This difference in style of presentation does not depend on the style of the research presented. Bieberbach was both a sloppy and suggestive author while Brouwer, on the other hand, wrote a formalist style of his own that resembles more that of a lawyer than of a mathematician.

For what Bieberbach had properly in mind when he considered differences of style of research I would like to propose terms like structuralism and realism rather than formalism and intuitionism. The structuralist *par excellence* and the father of modern structuralism would be Dedekind, the creator of such structuralist abstractions as field, ring, module and so on. Structuralism was to become the style of algebra and general topology. But — it is an old story — as things go on, such structuralistic abstractions are deemed to become part of our mathematical reality, which in turn afterwards falls victim to renewed structuralisation. This is an ever continuing process and what appears as an opposition of realism against structuralism is a kind of generation conflict. When I studied in Berlin I had the opportunity to learn the most modern mathematics, except with respect to algebra. This subject was taught by Issai Schur, a great mathematician, who could not acclimatise to the new structuralism initiated by Steinitz and welcomed by the young, in particular by the school of Emmy Noether. In Göttingen it was Courant who detested structuralism in analysis although it was cultivated by some of his best students. Only such great mathematicians as Siegel could afford to ignore structuralist tendencies completely, even from Dedekind onwards, and nevertheless create works of a genius. But on the other hand, as a more recent example, I guess without profiting of Grothendieck's extreme structuralism, Faltings would never have been able to solve the quite realistic problems of old standing — one of the most recent sensations in mathematical research.

Bieberbach's idea of mathematical style is meaningful but he himself did not grasp its proper meaning as he chose the wrong terms. Possibly at that time nobody was able to grasp it but meanwhile recent history has taught us new lessons. Anyway, when relating style to race or even nationality, he was badly mistaken. The German Dedekind had been a paragon of structuralism as the Frenchman Poincaré was just his realistic counterpart.

Let us leave this subject in order to take up another one. Probably in no field of research in Germany was the percentage of Jews as large as in mathematics. However, Mehrtens' figures can cause misunderstandings. He states that almost one third of teachers of mathematics in institutions of higher learning have eventually left service under the Third Reich. However, among the third there were political cases, "aryan" mathematicians married to women with some Jewish ancestry and, if I am not mistaken, even people with a weak Jewish spot in their own ancestry back to 1800 — Hasse seems to have been such a case, and grateful for having been exempted, he converted himself to nazism rather late. The shrinking number of students at German universities, mentioned by Mehrtens, was due to the accelerated militarisation of Germany: joining the army as an officer became the least expensive and most promising career.

Let me touch another point: the relatively successful resistance of German akademe to nazification. The majority of the German intelligentsia was conservative, monarchist, nationalist, or some mixture of these properties. I cannot tell whether they were so to a higher or lower degree than those in other countries as little as I would be able to compare degrees of racism and antisemitism in various countries. But it greatly helps understanding the course of history if one notes that (other than the Second) the First World War did not touch German territory. By fighting his war up to the bitter end, Hitler taught Germany a precious lesson. On the contrary the First World War ended with an armistice at a moment when German troops still occupied almost all of Belgium, and large parts of France and Russia. No doubt many were taken by surprise. The famous *Dolchstosslegende* said: rather than being defeated we were betrayed. Hitler, among others, capitalised on these feelings. Though despising Hitler's manners and his appeal to the masses, conservatives eventually accepted him, also as a guardian against communism. This halfhearted acceptation explains, among other causes, the remarkable resistance of akademe to nazification, signalised by Mehrtens.

At one point I would like to correct Mehrtens. The preposterous use of mathematical statistics for underpinning racist prejudice and other kinds of discrimination is no German invention. Eugenics goes as far back as Francis Galton, and with Cyril Burt, A.R. Jensen, and many others, it has preceded, coexisted with, and survived the Nazi variant of racism.

Mary Jo Nye

EXPLANATION AND CONVENTION IN NINETEENTH-CENTURY CHEMISTRY*

In his little book *The Structure of Chemistry* (1961), the English physical chemist E.F. Caldin suggested the need for a serious analysis of the methodology of chemistry. Such analysis would contribute to discussions of scientific method, he said, for example, whether theories are explanatory as well as instrumental and what may be the status of theoretical entities like atoms.[1] Aside from the specific question of the status of the atom, Caldin's general challenge still lies largely unanswered.[2] Philosophers and historians interested in these matters mostly

*I am grateful for comments from Zev Bechler, Herbert Mehrtens, Robert Nye, Steven Shapin, and Terry Shinn, aiding me to rethink arguments from an earlier version of this paper. I acknowledge with pleasure support for my research by the National Endowment for the Humanities and the University of Oklahoma Research Council.

1. E.F. Caldin, *The Structure of Chemistry in Relation to the Philosophy of Science* (London and New York, 1961); and W.F. Coulson, review of Caldin, *British Journal for the History of Science* 13 (1962) 184-185.

2. For example, A.J. Rocke, *Chemical Atomism in the Nineteenth Century: From Dalton to Cannizzaro* (Columbus, 1984); N. Fisher, "Avogadro, The Chemists, and Historians of Chemistry", *History of Science* 20 (1982) 77-102, 212-231; M. Gardner, "Realism and Instrumentalism in NineteenthCentury Atomism", *Philosophy of Science* 46 (1979) 1-34; W.H. Brock ed., *The Atomic Debates. Brodie and the Rejection of the Atomic Theory* (Leicester, 1967); D. Knight, *Atoms and Elements. A study of Theories of Matter in England in the Nineteenth Century* (London, 167); and M.J. Nye, "The Nineteenth-Century Atomic Debates and the Dilemma of an 'Indifferent Hypothesis'", *Studies in the History and Philosophy of Science* 7 (1976) 245-268. John H. Brooke and Alan Rocke have addressed issues other than atomism in the epistemology of chemistry: Brooke, "Laurent, Gerhardt, and the Philosophy of Chemistry", *Historical Studies in the Physical Sciences* 6 (1975) 405-429; and Rocke, "Hypothesis and Experiment in the Early Development of Kekulé's Benzene Theory", *Annals of Science* 42 (1985) 355-381. Hannah Gay has tested contrasting methodological prescriptions of Karl Popper and Imre Lakatos against the historical development of the radical and type theories in organic chemistry: Gay, "Radical and Types: A Critical Comparison of the Methodologies of Popper and Lakatos and Their Use in the Reconstruction of Some Nineteenth-Century Chemistry", *Studies in the History and Philosophy of Science* 7 (1976) 1-52.

have focussed on physics, biology (especially evolutionary biology), and the social sciences, with only occasional mention of the epistemology of atom-spheres, benzene hexagons, and the periodic table.[3]

By and large, a pejorative view of the methodological sophistication of chemical science has prevailed, notably in comparison to physics. On the one hand, chemistry has been described as a dispenser of recipes or the activity of "faire la cuisine."[4] Indeed, in the late nineteenth century, the French chemists Marcellin Berthelot and Henri Sainte-Claire Deville contributed to the view of their *métier* as a descriptive, empirical science rooted in natural history. Recalling a chemistry course taught in this tradition, Robert Lespieau, of the next generation, remarked (in 1913):

> "Four and a half hours was all the time devoted to generalities; if one had doubled this time, it would not have been detrimental to the seventeenth property of chlorous anhydride."[5]

Skirting the Scylla of naive empiricism, chemistry has not escaped the Charybdis of naive realism. John Dalton's atomism is in fact unrelentingly and naively realist. His illustrative plates in the *New System of Chemistry* (1808) are claimed there to exhibit clearly "the mode of combination of some of the more simple cases" of ultimate particles forming bigger ones. Dalton's compatriot Alexander Williamson defended realism, albeit a more sophisticated variety, at a Chemical Society debate in 1869.[6] Among recent philosophers, John Bradley has claimed that many nineteenth-century chemists were naive realists:

> "To Avogadro and Cannizzaro, as to Couper and Kekulé, the molecules and atoms

3. Important exceptions to disinterest in chemistry come from the French tradition, for example, in the work of Gaston Bachelard and François Dagognet: Bachelard, *La pluralisme cohérent de la chimie moderne* (Paris, 1932) and Dagognet, *Tableaux et langage de la chimie* (Paris, 1969).

4. In his *Confessions d'un chimiste ordinaire* (Paris, 1981), Jean Jacques mused that chemists rarely write for the public like their biologist and physicist colleagues. "Chemistry", he said, is "like the maid occupied with daily civilization: she is busy with fertilizers, medicines, glass, insecticides ... for which she dispenses the recipes" (p. 5). The physicist Henri Bouasse enraged his Nobel-Prize winning colleague Paul Sabatier by claiming that chemists only "faire la cuisine." On this, see M.J. Nye, *Science in the Provinces. Scientific Communities and Provincial Leadership in France, 1860-1930* (Berkeley, 1986) 286, n. 100.

5. R. Lespieau, "Sur les notations chimiques", *Revue du Mois* 16 (1913) 257-278, on p. 259. Also, M.J. Nye, "Berthelot's Anti-Atomism: A 'Matter of Taste'?", *Annals of Science* 38 (1981) 585-590.

6. J. Dalton, *A New System of Chemistry* (Manchester, 1808). On Williamson, see Brock, *Atomic Debates*.

considered in this great theory were real objects: they were thought of the same way as one thinks of tables and chairs."[7]

François Dagognet also has associated nineteenth-century chemical theory with "the pictorial hope of realism" and suggested that the *fin-de-siècle* was the dawn of reckoning for such epistemology.[8]

Chemists themselves, both now and in the nineteenth century, have not been as prone as other scientists to systematic reflection upon the methodology of their science. However, we can reconstruct many of their views by looking carefully at lectures, journal articles, and textbooks which were often cited or generally acknowledged to be influential during the nineteenth century. The metaphorical language and pictorial imagery of these texts can be misleading, especially upon a first reading and with the hindsights of modern physics and chemistry. However, a more careful reading of context enables us to understand the aims of non-mathematical and non-descriptive imagery and the conditions of its use.

As E.F. Caldin notes in *The Structure of Chemistry*, there are two kinds of empirical law in chemistry: 1) laws defining "natural kinds", similar to biological classification; and 2) laws stating functional relations between variable properties.[9] Both in establishing these laws and in articulating theories interpreting them, nineteenth-century chemists self-consciously negotiated working definitions and postulates as systematic conventions for their discipline. More so than natural philosophers or physicists, I think, they mostly regarded their theories or explanations as provisory and contingent, and based on conventions. Accordingly, they were not hesitant to employ logically contradictory hypotheses or images in the service of better prediction and explanation in chemical synthesis. I.e., they were not obsessed with logical rigor.

However, this does not mean that nineteenth-century chemists were dyed-in-the-wool instrumentalists or radical skeptics, any more than they were naive empiricists or naive realists. My aim is to understand what they were doing, suggest how chemical explanation differed from physical explanation, and to demonstrate in conclusion the extent of the practical application of conventions in nineteenth-century chemistry. To these ends, I shall analyse chemical explanation, mostly in organic chemistry, in terms of three categories, which I will call: 1) the conventionalization of descriptive metaphors; 2) the development of symbolic, non-visual formulae, superficially similar to algebra; and 3) the

7. J. Bradley, "On the Operational Interaction of Classical Chemistry", *British Journal of the Philosophy of Science* 6 (1955-1956) 32-42, on p. 32.
8. Dagognet, *Tableaux*, 199.
9. Caldin, *Structure*, 18-19, 26, 28-29, 31-32.

formulation of schematic, visual formulae, superficially suggestive of physical modelling. Such devices, we shall see, lent themselves to representation of the multiple *functions* of a chemical molecule, rather than representation of three-dimensional reality, as the aim of chemical explanation.

Let us begin with some fundamental conventions of nineteenth-century chemistry. Historians of chemistry have shown how a new chemical language and nomenclature became vehicles of institutional and conceptual reform in the work of Antoine Lavoisier, Guyton de Morveau and their colleagues on the eve of the nineteenth century.[10] The reformed system was explicitly conventional and chemists were persuaded to adopt it. Yet despite its Enlightenment nomenclature, chemistry remained a language of replete and redolent metaphor, often immune to the "mathematical plainnesse" thought appropriate to physical science.[11]

Consider some of the enduring and legion examples of such devices. In the mid-nineteenth-century chemistry of Auguste Laurent, atoms or particles were active in the "chase", in "copulation", "conjugation", "birth", and "marriages of convenience".[12]
A.W. Hofmann's aldehydes absorbed oxygen not indifferently, but with "greed".[13] For Henry Armstrong, molecules were keys fitting locks; ducks on a ramrod; and curls, like a judge's wig.[14] Jean-Baptiste Dumas in the 1830s had revived Leucippus' trope that particles of matter are "like letters transposed, which thus form altogether different words".[15] For Laurent, particles form structures like Gothic cathedrals:

> "Some are ornated with two towers, others have only one, some are topped with

10. For example, M. Crosland, *Historical Studies in the Language of Chemistry* (Cambridge, 1962); H. Guerlac, *Antoine-Laurent Lavoisier: Chemist and Revolutionary* (New York, 1968); D. McKie, *Antoine Lavoisier: Scientist, Economist, Social Reformer* (New York, 1952); and most recently, from a literary analysis, W.C. Anderson, *Between the Library and the Laboratory. The Language of Chemistry in Eighteenth-Century France* (Baltimore, 1984).

11. See Th. Sprat, *History of the Royal Society*, 111, 113, quoted in W.H. Leatherdale, *The Role of Analogy, Model and Metaphor in Science* (Amsterdam, 1974) 223-224.

12. A. Laurent, *Méthode de chimie* (Paris, 1854) 29.

13. A.W. Hofmann, "On the Combining Power of Atoms", *Chemical News* 12 (1865) 166-169, 175-179, and 187-190, on p. 166.

14. H.E. Armstrong, "Presidential Address of the Chemical Section", *Reports of the British Association for the Advancement of Science* (Winnipeg, 1909) 420-454, on p. 443-446.

15. J.B. Dumas, *Leçons sur la philosophie chimique* (Paris, [1837]. Deposited at the Bibliothèque Nationale, 1840) 316.

steeples, with or without lateral galleries, but the fundamental plan of the edifice is always the same, it is the Greek cross."[16]

Or chemical form became biological in Laurent's metaphor of the chemical "tree":

"I have searched if there does not exist in all parts of a same chemical tree, something analogous to this mother cell, in a word a nucleus common to all compounds of the same series."[17]

Like so many chemists, Armstrong also could not resist the military metaphor, for example, in describing the chemistry of camphor:

"whatever the agent, the attack is always delivered from the oxygen center and ... the direction in which the attack becomes effective depends on the position which the agent can take up relatively to the various sections of the molecule."[18]

But this metaphorical, descriptive language should not be misleading. It is not naive. It is of a whimsical, not serious nature. All my examples, with the exception of the "greedy" aldehydes in Hofmann's public lecture, originate in texts aimed at fellow chemists, intended to be illuminating, but not meant to be *rigorous*. However, in contrast to such whimsical and playful description, some other anthropomorphic or ordinary language metaphors developed into serious definitions, or conventions, which were essential to rigorous chemical classification and explanation. For example, consider the following terms: affinity, atomicity, (equi)valence.

The term affinity is rooted in Newton's notion of force. The affinities appearing in E.F. Geoffroy's table of 1718 were understood to be analogous to Newton's gravitational forces.[19] But if affinity was *like* gravity, electricity, or, to a lesser extent, magnetism, it was not identical to them. An important difference was that "affinity" was more selective or "elective" than its analogues. Defining affinity in 1867, Adolphe Wurtz called it the "force which presides over chemical combination,"[20] a metaphor of governance similar to "Gravity ruling" or "Nature selecting." In his last years, Wurtz wrote pensively:

"We have thought of affinity or chemical force as an attractive force, like a form of universal gravitational attraction exerted between celestial bodies. It is a hypothesis

16. Laurent, *Méthode*, 176.
17. *Ibid.*, 397-398.
18. Armstrong, "Address", 440.
19. A. Thackray, *Atoms and Powers. An Essay on Newtonian Matter-Theory and the Development of Chemistry* (Cambridge, 1970) 90, 213. For an early history of affinity, see Dumas, *Leçons*, 359-392.
20. A. Wurtz, *Leçons élémentaires de chimie moderne* (Paris, 1867) 6.

> in both cases, or if one wants, a useful representation. It is figurative language. It is a way of expressing a fact rather than giving an explanation."[21]

Thus, as in the physicists Heinrich Hertz's and Ernst Mach's late nineteenth-century critiques of "force", Wurtz recognized "affinity" to be a definition, a convenient figure of speech for talking about chemical change and its properties.

In a second example, consider the word "atomicity". Wurtz in 1867 defined "atomicity" as the "force or power of combination which resides in atoms, and which is exercized in different manner according to the nature of the atoms."[22] However, the existence of indivisible particles corresponding to the chemical elements was itself in dispute among chemists. Many used the word as a convention, without commitment to its etymological meaning. Thus, Hofmann wrote in his *Introduction to Modern Chemistry* (1865) that "atomicity" was a barbarous expression because it suggested atomic structure. He preferred the word "quanti-valence", a word shortened to valence (*Valenz*) by Auguste Kekulé and then Hermann Wichelhaus.[23] The valence, or value, of an element varies according to the number of hydrogen atoms with which it can combine.

Another word expressing this idea was "saturation" capacity. Using an electrochemical theory of chemical reaction, J.J. Berzelius hid the assumption of electrical neutralization by the material analogy or metaphor of saturation.[24] "Valence" was meant to be a more neutral word than "atomicity" or "saturation capacity", originating from the term "equivalence" which, without linguistic implications of atoms, expressed exact combining proportions of weights of substances with one another. Thus, using hydrogen or oxygen as a standard substance, and assigning the number 1, 10, or 100 as the standard dimensionless unit for hydrogen or oxygen, tables of equivalents could be constructed.

However, chemists soon recognized that both systems were based in convention. The atomist Berzelius wrote Justus von Liebig to this effect, that the system of equivalents was "no longer positive, but merely conventional".[25] Typically, when Dumas explained the system to his students in 1837, he frequently used the words "convention" and "convenir", as did Liebig in his introductory lectures in chemistry.[26]

21. A. Wurtz, *Introduction à l'étude de la chimie* (Paris, 1885) 8-9.
22. Wurtz, *Leçons élémentaires*, 219.
23. See C.A. Russell, *A History of Valency* (New York, 1971) 85-86.
24. *Ibid.*, 23.
25. Quoted in Rocke, *Chemical Atomism*, 206.
26. Dumas, *Leçons* 225-228; J. von Liebig, *Introduction à l'étude de la chimie*, trans. Ch. Gerhardt (Paris, 1837) 85.

Most chemical atomists, like Thomas Thomson in 1813, made few truth claims. Thomson submitted his table of atomic weights, for example, "to the chemical world as more convenient".[27] Conventions, then, were at the base of nineteenth-century chemistry and widely recognized to be so. This fact contributed to the reticence of chemists in making confident claims about the correspondence of their systems with reality.

Let us turn now from these kinds of conventions to the use of symbolic formulae which were superficially similar to algebraic equations. Lavoisier explicitly warned that his occasional symbols were not like algebraic symbols; rather, they were heuristic instruments or shortcuts for thinking.[28] Whereas Dalton's table of atomic weights included geometrical symbols for the elements and compounds, and Berzelius, like Dalton, imagined elemental atoms as spheres, Berzelius broke with pictorial or geometrical symbols. Instead, he used letters of the alphabet, index numbers, dots, bars, and the sign of addition to represent so-called double-decomposition reactions (electrochemical dualism).[29] Very quickly these alphabetical linear formulae included grouped letters within the linear formulae, and Berzelius introduced a distinction between "empirical" and "rational"formulae, the latter indicating how atoms are grouped together within a molecule or compound.[30] Thus:

$$NH_3 \, . \, HCl \text{ or } KO.NO_2$$

In the 1830s, the so-called organic "radical" theory began, with the announcement by Liebig and Friedrich Wöhler that some atoms form groupings which remain units under different circumstances of reactions, for example, the benzoyl "radical" or root, composed of carbon and oxygen (now, C=O).[31] It was easy to represent these "radicals" by letters and subscripts separated spatially from other letters representing

27. Th. Thomson, "On the Daltonian Theory of Definite Proportions in Chemical Combinations", *Annals of Philosophy* 2 (1813) 32-43, on p. 41-42.

28. Included in the 1787 collection of memoirs making up the *Méthode de nomenclature chimique* were six folding plates demonstrating a symbolic scheme for chemical representation devised by Lavoisier's colleagues Hassenfratz and Adet. For a discussion, see Crosland, *Historical Studies*, 240-248.

29. On Dalton's symbols, see W.W. Haldane Gee, H.F. Coward and A. Harden, "John Dalton's Lectures and Lecture Illustrations", Memoir no. XII in *Memoirs and Proceedings of the Manchester Literary and Philosophical Society* 59 (1914-1915), esp. p. 41-66. Also, Crosland, *Historical Studies*, 268-275.

30. *Ibid.*, 323.

31. *Ibid.*, 302; and O.Th. Benfey, *From Vital Force to Structural Formulas* (Washington, 1975; 1st ed., 1964) 34.

elements in a compound. However, from the start, chemists debated whether the "radicals" were groups "convenient for description" or real organic equivalents of inorganic elements. Dumas claimed that radicals really exist, that NH_3, for example, is "preformed" in ammonium chloride. Laurent disagreed, criticized the inclusion of over one-hundred hypothetical radicals in one of Liebig's treatises, and argued that preformed groups could not be inferred from reactions because atoms may be arranged differently in a non-reacting compound than in a reacting one.[32] Kekulé shared Laurent's view, as did Laurent's colleague and ally Charles Gerhardt, who rechristened the radicals "residues" and insisted that the units were arbitrarily chosen for the convenience of the chemist.[33] Whereas Berzelius had seemed to have a Platonic notion of one "true" rational formula, Gerhardt's motto was "autant de réactions autant de formules rationelles."[34]

The formulae began to take on a non-linear presentation both in the electrochemical and the radical theories. In 1850 the English chemist Thomas Graham suggested writing a formula in two lines, placing the negative constituents on the upper, and the positive on the lower line. Such formulae, he suggested, would "exhibit" the polar attractive forces of the elements.[35] Thus:

$$\text{potash}\ \frac{O}{K} \qquad \text{ammonia}\ \frac{N}{H_3} \qquad \text{and} \qquad \frac{N}{H_3} \qquad \text{and} \qquad \frac{Cl}{H}\ \text{give} \qquad \frac{NCl}{H_2H_2}$$

The new "type" theory similarly shifted to non-linear formulae, reminding the reader of the crystallographic and biological *forms* which were the inspiration for the theories of water, ammonia, hydrogen-chloride, and methane "types". Gerhardt used elongated brackets in 1856, as did then Edward Frankland and William Odling, although Williamson would not employ them. Gerhardt insisted that the arrangement of brackets, letters, and other marks on paper did not imply actual twodimensional, physical arrangement of parts.[36] Thus, the water "type" and other variations:

$$\left.\begin{matrix} H \\ H \end{matrix}\right\} O \qquad \left.\begin{matrix} K \\ K \end{matrix}\right\} O \qquad \begin{matrix} H \\ SO_2 \\ H \end{matrix} \left\{\begin{matrix} O \\ O \end{matrix}\right.$$

32. Brooke, "Laurent", 411-412, 428.
33. Benfey, *Vital Force*, 58.
34. Crosland, *Historical Studies*, 331.
35. Th. Graham, *Elements of Chemistry: Including the Application of the Science in the Arts*, 2nd ed., 2 vols (London, 1850) I, 205-207.
36. Russell, *Valency*, 94-95.

And, from Wurtz,[37] (note the grappling hook):

With the exception of some unique symbols of William Higgins in 1789, by and large, straight lines appeared in published chemical formulae only when Archibald Couper introduced them in 1858 to indicate valences (atomicity, saturation capacity, or quantivalence).[38] Whereas innocent accent marks or superscript dashes had been used at mid-century to indicate valence, straight lines suggested a less abstract meaning, despite disclaimers like that of Alexander Crum Brown that the lines indicated the "chemical", not "physical", positions of atoms.[39]

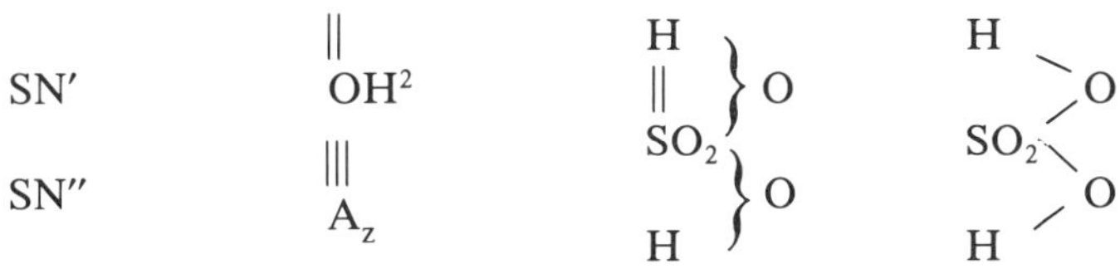

As Colin Russell has noted in his *History of Valency* (1971), a straight line is a much "more tangible symbol ... like rods, tiebars, and so on."[40] The pictorial or structural formula had made its debut.

In the first half of the nineteenth century, chemists continually discussed among themselves the status of these early symbols and formulae.[41] Chemical symbolism was a topic at the first meeting in 1834 of the chemistry section of the British Association for the Advancement of Science. This followed by three years William Whewell's critique of chemical symbols in the *Journal of the Royal Institution*.

> "Its formulae are merely unconnected records of inferences which are in some degree arbitrary; the analysis itself, the fundamental and certain fact from which inferences are made, is not recorded in the symbol; and the connection between different formulae, the identity of which is a necessary and important circumstance, can be recognized only by an entire perversion of all algebraical rules."[42]

37. Wurtz, *Leçons élémentaires*, 77.
38. Russell, *Valency*, 100-101.
39. *Ibid.*, 102.
40. *Ibid.*, 100.
41. Armstrong, "Address", 427.
42. See R.F. Bud and G.K. Roberts, *Science versus Practice. Chemistry in Victorian Britain* (Manchester, 1984) 39-40, 44.

Although Whewell worked out a system of analytical statements for chemists, they did not use it.[43] Benjamin Brodie renewed the project for algebraic reform of chemical formulae in a paper presented to the London Chemical Society in 1867. But again, there was little interest.[44] Chemists did not find this kind of algebraic representation with Greek symbols useful, in contrast to the increasingly visual imagery in formulae using the chemical bond, in clearly non-algebraic formulae.

Let us look now at some of the best-known examples of the visual formulae. For example, the chemistry of benzene and its derivatives elicited the publication of Kekulés famous hexagon formula in 1865,

followed by Ladenburg's non-planar

prism formula.[45] Alexander Butlerov and then Kekulé proposed the carbon-valence tetrahedron, an idea further developed by J.A. LeBel and J.H. Van 't Hoff. The so-called "unsaturated" properties of benzene and other hydrocarbons, their ability to "soak up" more hydrogen, for example, resulted in cartoon-like graphical representations for the valence bond. Thus,:[46]

43. *Ibid.*, and Crosland, *Historical Studies*, 278-279. The resistance of some French chemists to the "constitution" ("developed" or "structural") formulas became notorious. As late as 1890, Henri Moissan published a fifteen-page memoir on ethyl fluoride in which no formula appears. In 1891, Berthelot stopped using equivalents, but he never used constitution formulas. The ninth edition of his book on chemical synthesis, appearing in 1903, contained not one change in notation from the 1876 edition. See the discussion by one of France's leading twentieth-century organic chemists, A. Kirrmann, "La naissance des formules moléculaires en chimie organique", *L'Actualité chimique* (janvier 1974, no. 1) 45-49, esp. p. 48. Kolbe behaved similarly in Germany.

44. B. Brodie, "On the Mode of Representation afforded by the Chemical Calculus, as Contrasted with the Atomic Theory", *Chemical News* 15 (1867) 295-305, esp. p. 298-300. This and other seminal chemical papers of the nineteenth century are reprinted in D.M. Knight ed., *Classical Scientific Papers Chemistry* (New York, 1968). Also, see Brock, *Atomic Debates*.

45. Russell, *Valency*, 243 ff.

46. *Ibid.*, 234; and Benfey, *Vital Force*, 105-107.

Alexander Crum Brown (1864)

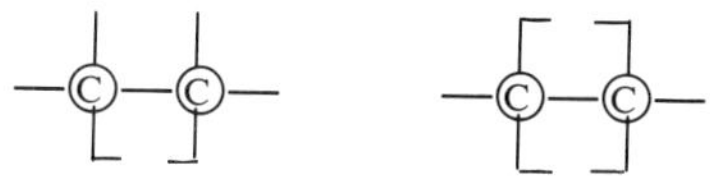

Josef Willbrand (1865)

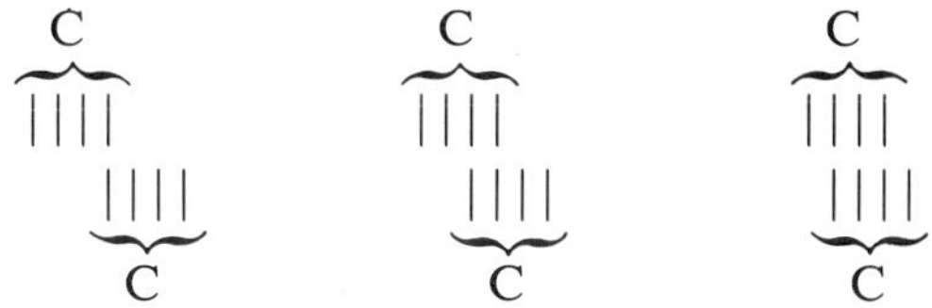

Kekulé (1861)

The puzzling behavior of benzene, its unreactivity and the inference that all six carbon atoms must be linked in the same way, could not be described by any one image or cartoon.[47] Rival schemes included figures in which Josef Loschmidt used a large circle to signify neither atoms nor a ring structure for benzene, but a structurally indeterminate benzene nucleus.[48]

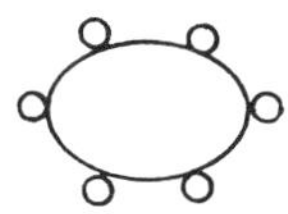

A truly impressionistic, or *pointillist*, scheme appears in the work of Johannes Thiele, who attempted to explain anomalies in the behavior of butadiene, as well as benzene (cyclohexatriene). Thus, we have Thiele's "floating" or "indeterminate" valences which remained in use in amended form until the end of the 1920s.[49]

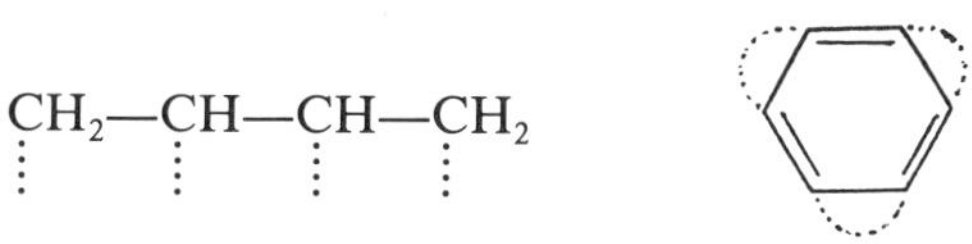

47. Ch. Friedel, *Cours de chimie organique* (Paris, 1887) 310-311; and L. Meyer, *Die modernen Theorien der Chemie und ihre Bedeutung für chemische Statik*, 3rd ed. (Breslau, 1877) 195.

48. Rocke, "Hypothesis and Experiment", 366.

49. F.K.J. Thiele, "Zur Kenntnis der unsättigten Verbindungen. Pt. I.", *Annalen der Chemie* 306 (1899) 87-142, on pp. 89, 126. Also, Russell, *Valency,* 257; and Dagognet, *Tableaux*, 200-201.

However, it is clear that, on the whole, those chemists who developed and used these representations had no intention for them to be understood as reflections of two- or three-dimensional reality. Nor, by and large, did they give them the status of theoretical models.[50]

To be sure, most chemists were convinced by the existence of isomers and optical rotatory power that "molecules" exist as real, three-dimensional structures of some sort.[51] But their schemata of representation were heuristic, often short-lived, and intended to be so. In fact Kekulé himself rarely used the hexagonal structure in papers,[52] and his controversial "sausage" or "roll" diagrams for hydrocarbons appeared only in two published works and in lectures notes of 1857-1858.[53]

What then shall we conclude about the meaning of the pictorial formulae and schemata? What did they explain? Why were they not theoretical models? Benjamin Brodie amused his audience at the Chemical Society in 1867 with a scathing description of a commercially available set of molecular models:

> "There are seventy balls in all for the representation of atoms monads, dyads, triads, tetrads, pentads, and hexads [valence] being distinguished by the number of holes pierced in the balls. To connect these into rational formulae ... brass rods, straight or bent, and occasionally flexible bands are employed."[54]

We see that Brodie here clearly interchanges the roles in chemical explanation of the rational or structural formula and the molecular model, because the rest of his lecture is a critique of the formula. But were the ball-and-rod devices "models" in the usual sense of explanation in the physical sciences? What was the aim of A.W. Hofmann's croquet balls, Dewar's brass-and-rod models, Kekulé's wooden sausages, and Van 't Hoff's cardboard tetrahedrons?[55]

In fact, these were not like orreys, devices imitating the structure and movements of the universe. Because they were visual images, the

50. An important exception appears to be A.W. Williamson. See Williamson, "On the Constitution of Salts", *Chemical Gazette* 9 (1851) 334-339. This and other papers are reprinted in O.Th. Benfey ed., *Classics in the Theory of Chemical Combination* (New York, 1963).

51. See J.H. Van 't Hoff, "Die Lagerung der Atome im Raume", trans. F. Herrmans (Braunschweig, 1877) 10-11, 17.

52. Rocke, "Hypothesis and Experiment", 371, n. 81.

53. Russell, *Valency*, 98-100.

54. Brodie, "Chemical Calculus", 296.

55. Rocke, "Hypothesis and Experiment"; J.R. Partington, *A History of Chemistry* (London, 1964) IV, 556; Crosland, *Historical Studies*, 336. See also B. Ramsay, "Molecules in Three Dimensions", *Chemistry* 47, no. 1 (1974) 6-9, and no. 2 (1974) 6-11; and, Ramsay, *Stereochemistry* (London, 1981).

developed rational formulae and later stereochemical images of nineteenth-century chemistry seem easily transformable to the model-as-orrey idea. But chemists who employed them denied that their schema really reflected nature as it is. For example, Van 't Hoff said, "The representations themselves, atom, molecule, their dimensions, and perhaps their shapes, are after all something doubtful, as is the tetrahedron itself."[56]

In addition, the role of a model as a mediating link between mathematical expression and palpable experience was inapplicable, because at this time in organic chemistry, the formulae or representations were the endpoint of the theory, leading directly into prediction or explanation through an understanding of function. In short, the explanatory images or visual metaphors did not fulfill the chemists' own requirements for a model. Chemists regularly understood models to involve mechanisms, namely dynamic relationships, and their images were only beginning to suggest mechanics at the end of the century.

For example, reflecting in 1929 on the function of models in science, the chemist Georges Urbain described a model as an ideal image which must obey the same laws as perceived facts. It is a "variation of the procedure of analogy covered by the name of metaphor." Explanatory hypothesis, he claimed, always takes the form of a model.[57] But chemical representations in the nineteenth century, in Urbain's view, were not true models; and for Urbain, writing from the twentieth-century perspective, they were not clear explanations. The double tetrahedron, by which the double bond is represented by one of the sides, is "a pure geometrical symbol", because

> "... it is unconcerned with the forces issuing from the atom considered to be a compact and homogeneous block at the center of the tetrahedron ... it hardly conforms to modern scientific thought to confuse purely symbolic figures which are synoptic signs of writing with mechanisms which are fictive phenomena substituted for the real phenomena whose laws they share."[58]

When Archibald Couper used lines or bars in graphic formulae, he did have in mind a representation of the forces of attraction between the elements in organic molecules.[59] Lothar Meyer and Kekulé were among those who quickly, but circumspectly adopted the notation. But Edward

56. Quoted in A.G. Van Melsen, *From Atoms to Atom* (New York, 1960) 151.

57. G. Urbain, "Remarques sur l'orientation des doctrines chimiques", *Revue de Métaphysique et de Morale* 36 (1929) 177-191, on p. 180-181.

58. G. Urbain, *La coordination des atomes dans la molécule. La symbolique chimique.* Actualités Scientifiques et Industrielles, nos. 52 and 53 (Paris, 1933) 13.

59. Crosland, *Historical Studies*, 331-332.

Frankland tried to be very careful in using the word "bond" and its bar representation:

> "By the term bond, I intend merely to give a more concrete expression to what has received various names from different chemists, such as atomicity, an atomic power, and an equivalent ... It is scarcely necessary to remark that by this term I do not intend to convey the idea of any material connection between the elements of a compound, the bonds actually holding the atoms of a chemical compound being, as regards their nature, much more like those which connect the members of our solar system."[60]

Logically, more bonds between atoms should mean a greater force of attraction and increased carbon density. But doubly-bonded alkenes are more reactive, not less reactive than singly-bonded alkanes or triply-bonded acetylenic compounds.[61] Further, how could the force associated with a polyvalent atom be divided *a priori* to react to the atoms encountered, sometimes dividing itself in half, other times by thirds? To assume that valences act across empty space, free of atoms, some chemists noted, "is only possible on paper or in a model where there are lines or wires but not forces."[62]

Most chemists were conscious that their molecular formulae were static representations of what must really be a phenomenon of dynamics. Even the "atom" itself likely had a substructure capable of dynamic modelling.[63] Gerhardt, Kekulé, and others were convinced that the molecule-at-rest must have a different internal structure than the molecule in reaction. The idea that atoms within the molecule are in constant motion received increasing influence after Williamson's 1850 paper on etherification, where he concluded that an exchange of atoms between molecules goes on all the time.[64] Further, it was discovered that some compounds behave as if they have two different structures. Peter Laar in 1886 suggested that this phenomenon of "tautomerism" is due to a continual oscillation of a hydrogen atom between two positions within a single molecule,[65] a hypothesis influenced by Kekulé's suggestion that

60. In Russell, *Valency*, 90. See E. Frankland, *Lecture Notes for Chemical Students: Embracing Mineral and Organic Chemistry* (London, 1866) 25.

61. Russell, *Valency*, 237.

62. Quoting R. Demuth and V. Meyer, in Russell, *Valency*, 166.

63. See Nye, "Nineteenth-Century Atomic Debates" and "Berthelot's Anti-Atomism"; and A. Rocke, "Subatomic Speculations and the Origin of Structure Theory", *Ambix* 30 (1983) 1-18.

64. A.W. Williamson, "Theory of Aetherification", *Philosophical Magazine* 37 (1850) 350-356, on p. 356; Graham, *Elements*, 600-603; and H. Watts ed., *A Dictionary of Chemistry and the Allied Branches of Other Sciences*, 5 vols, 2nd ed.: Watts' article on "Chemical Affinity" (London, 1872) I, 850-866, on p. 866.

65. A. Findlay, *A Hundred Years of Chemistry*, 2nd ed. (London, 1948; 1st ed. 1937) 119.

the peculiarities of benzene are due to an oscillation of atoms in benzene. Kekulé's specific hypothesis was that "equivalence" or valence is "the relative number of contacts which occur in the unit of time between atoms."[66]

Thus, the inner workings of things would be explained by a mathematical model, perhaps like the vortex theory of the atom. In the meantime, the *function* of things had been worked out by chemists' analogical reasoning, expressed in increasingly graphic and pictorial imagery. These images portrayed alternative functions of single substances under different conditions, according to conventions agreed upon by most chemists. A set of complementary representations of a hydrocarbon might emphasize first its acid character, now its aldehyde property, then its substitution behavior. This was a great achievement of nineteenth-century organic chemistry.

The French anti-Berthelot school was clear on this point, beginning with Laurent, who wrote that "The formula represents the functions of the compound,"[67] a point of view shared by Wurtz and Edouard Grimaux. Wurtz claimed that "These formulas ... do not give any indication on the form of the molecule in space."[68] Similarly, Grimaux said:

> "... none of us imagine, with figures traced in a plane, to determine the real place of atoms in space ... these formulas recall to us the reactions themselves."[69]

In this tradition, Charles Friedel argued at the 1892 international Geneva congress on nomenclature against the German demand for one official name for each compound. In contrast, the French chemists wanted only rules for arriving at the several different names that are equally valid in describing each particular function of a substance.[70]

In conclusion, it is clear that many nineteenth-century chemists were neither militant empiricists nor naive realists. And the successes of twentieth-century applications of physics to chemistry, as in x-ray crystallography, should not distort our understanding of how nineteenth-century chemists proceeded and what they aimed to explain. In the end, their explanations of chemical *function* did provide a key to molecular

66. *Ibid.*, 125; and A. Kekulé, "On Some Condensation Products of Aldehyde", *Annalen der Chemie und Pharmacie* 162 (1872) 77-124, abstracted by H. Armstrong in *Journal of the Chemical Society, Abstracts* 25 (1872) 612-619, on p. 614.

67. Laurent, *Méthode*, 421.

68. Wurtz, *Introduction*, 255.

69. E. Grimaux, *Introduction à l'étude de la chimie. Théories et notations chimiques* (Paris, 1883) 145.

70. Crosland, *Historical Studies*, 347-348.

form, and for that matter, to electron distributions in space. This result would neither surprise nor chagrin most nineteenth-century chemists specializing in organic reactions. For if they characterized their knowledge as provisory and contingent, conventional at base, and susceptible to multiple representations, they were nonetheless confident that their science was progressive and faithful to the phenomena.

There are significant comparisons to be made here with the so-called conventionalist epistemology of Henri Poincaré, which attracted a large audience in the early 1900s, including one among chemists. Poincaré repudiated both the instrumentalism of Pierre Duhem and the radical relativism or nominalism of Edouard LeRoy. As Jerzy Giedymin has noted in his studies of Poincaré, Poincaré abandoned the Kantian notion of *a priori* intuitions as the basis for scientific theories, in favor of negotiated conventions.[71] But Poincaré was a progressivist, too. He acknowledged the creative role of scientific hypotheses and conventions without succumbing to the notion that hypotheses and conventions are arbitrary; he stressed that they are rooted in real phenomena. The historical record might be one of theories abandoned one after another, but for Poincaré, that record does not deny rational and progressive *enchaînement* over time.[72] The same may be said for many nineteenth-century chemists, especially in organic chemistry.

The lack of mechanical models and rigorous mathematics makes nineteenth-century chemistry a different science from physics, but no less methodologically sophisticated, as we have seen. Many chemists employed metaphors, images, and conventions with selfconscious examination and debate among themselves. The grand old man of British chemistry Henry Armstrong did not quail before the non-mathematical and non-rigorous character of his science at the end of the nineteenth century. Commenting on the remark made by the physicist Hermann von Helmholtz that organic chemistry progresses "not quite rationally", Armstrong riposted that this remark was little more than a confession that Helmholtz "was out of his depth".[73]

University of Oklahoma

71. J. Giedymin, *Science and Convention. Essays on Henri Poincaré's Philosophy of Science and the Conventionalist Tradition* (Oxford, 1982) VIII; and M.J. Nye, "The Boutroux Circle and the Origins of Poincaré's Conventionalism", *Journal of the History of Ideas* 40 (1979) 107-120.

72. See, for example, H. Poincaré, "Sur la valeur objective de la science", *Revue de Métaphysique et de Morale* 10 (1902) 283, 285-286.

73. Armstrong, "Address" 423.

COMMENTARY

Willem J. Hornix *On the paper of Mary Jo Nye*

Mary Jo Nye offers an analysis of what 19th-century chemists understood by "chemical explanation". Her conclusion is already given at the start of her paper, when she describes the theoretical work of these chemists as an activity in which they "self-consciously negotiated working definitions and postulates as systematic conventions for their discipline ..." and "regarded their theories or explanations as provisory and contingent, and based on conventions".

Her analysis is structured through the exposition of three categories which I understand as characterizing the theoretical development of 19th-century organic chemistry. The first category describes a process starting from descriptive metaphors and leading to conventional definitions. Her examples are the concepts of affinity and valency.

The second and third categories characterize the development of the symbolic formulae for chemical compounds from pictorial symbols, which are shortcuts for thinking, via the alphabetical linear formulae of Berzelius, towards the non-linear visual formulae of Gerhardt and Kekulé.

In the end these formulae expressed nothing more than the chemical function of the molecule, in other words the pattern of reactions of the chemical compounds associated with the structural formulae, and consequently these are purely conventional.

I will comment on Professor Nye's paper by consecutively considering the development of the concepts "affinity," "equivalence," "valency," and "chemical formula".

Professor Nye's opening statement about affinity, that the term is "rooted in Newton's notion of force," neglects its use by chemists before Newton. The 18th-century affinity tables describe replacement-reactions. "Elective affinity" expresses precise ordering relations and although a measure is missing the concept pretends to cover an important domain of experimental results.[1]

1. See A.M. Duncan, "Some theoretical aspects of eighteenth-century tables of affinity I", *Annals of Science* 18 (1962) 181, and R.E. Schofield, *Mechanism and Materialism, British Natural Philosophy in an Age of Reason* (Princeton, 1970) 216.

Berthollet's treatment of affinity in his *Recherches sur les lois de l'affinité* (1799) is also purely conventional. He succeeded in removing the most conspicuous discrepancies between the older theory and experiment through the introduction of the factor "mass" into the theory; however, his attempt to define a measure for affinity ended in the recognition of its indeterminateness, i.e. the impossibility to determine the affinity of A for B quantitatively. Certainly, the 18th and 19th-century chemists attempted to explain the conventional affinity concept in a newtonian theory of some sort. Berthollet's *Essai de Statique Chimique* is only one example. Affinity is not a metaphor derived from newtonian attractions.

The discovery of the equivalence of work and heat offered the prospect of reducing affinity via its supposed measure — heat of reaction — to "internal work" of the newtonian attraction forces. This prospect turned out to be a mirage. And Wurtz's 1885 admission that affinity is a convenient figure of speech for talking about chemical change, presumably marks the failure of thermochemistry as an explanatory theory for the phenomena described through "affinity". Perhaps it also marks the recognition that the theory which offered at last a measure for affinity completely in agreement with experiment, phenomenological thermodynamics, is in fact the conventional theory par excellence for physicists and chemists alike.

The dynamics of the development of the affinity concept in the 18th and 19th centuries is the continuing sophistication of the conventional approach accompanied by endeavours to give a mechanical interpretation, until the ultimate success of conventional theory in thermodynamics and the persistent failure of the mechanical explanation.

The next examples are the related concepts of equivalence and valency or atomicity. Equivalence weights as introduced by Richter (1792) and Fischer (1802) describe double decomposition reactions of salts and neutralization reactions of acids and bases quantitatively, and are purely conventional. It was through Dalton's atomic theory that the system of equivalent weights was connected with atomic weights and extended to the elements. The equivalent and atomic weights of the elements are sets of numbers regulating the composition of all chemical compounds. Atomic weights — which belong to the realistic world of atoms and attractions — could only be determined with the help of extra presuppositions which were susceptible to strong criticism. Consequently many chemists went back to equivalent weights for the elements, accepting their relative arbitrariness, to gain a clear and unambiguous representation of the quantitative elementary composition of the known compounds in the "empirical formulae". This relative arbitrariness called for directions to guide their selection. Laurent (1854) chose:"Those that comply with the two following conditions: (1) to represent the different series of compounds by the simplest possible formulae, (2) to assign to analogous compounds analogous formulae."[2] Analogous compounds have to be understood as *chemically* analogous: the formulae derived for the compounds with the chosen equivalent weights have to express their chemical relationships. Laurent's choice is determined by convenience. Laurent, however, did not reject atomic weights and a realistic interpretation of equivalence weights. He introduced our modern definition of atom and molecule and provided selection criteria for atomic weights:

> "Since it is our object to discover the weight of particles, we shall consider that we have attained it if for each substance, simple or compound, we find a proportional number concordant with the form, the volume, and the specific heat of the substance; and if this number allows us to represent in a very simple manner the reactions and the formulae ..."[3]

One might conclude that because of the extra criteria, atomic weights are even more convenient than the equivalent weights. But considering these extra criteria we must keep in mind that isomorphism, volume relationships and specific heats have always been related to constitution, number and specific properties of atoms and molecules. The concordance pursued resolved difficulties standing in the way of the atomistic interpretation of the conventional theory. That it took another fifteen years before the proposals of Laurent and his ally Gerhardt were accepted, falls outside the scope of this comment.

The dynamics of the development can be summarized as follows: the chemical equivalent was introduced as a purely conventional concept, the extension of its domain of applicability was induced by Dalton's atomic theory, discrepancies of the consequences of the atomic theory and experience and the indeterminateness of atomic weights, caused a return to conventionalism. Further sophistication of the conventional theory finally brought it into complete accordance with an atomic interpretation.

Let us now consider the concept "valency" or "atomicity". Professor Nye introduces the concept by referring to Wurtz's definition of "atomicity" as a specific property of atoms. This is also the way in which the concept is still taught to our chemistry students. The historical origin is, however, different. Frankland, who ranks as its discoverer, remarked

2. A. Laurent, *Méthode de Chimie* (Paris, 1854) as quoted by I. Freund, *The Study of Chemical Composition* (Cambridge, 1904; Dover ed., 1968) 196.

3. A. Laurent, "Sur les combinaisons azotées", *Annales de Chimie et de Physique* 18 (1846) 289 as quoted by Freund, *Study*, 343.

that "even a superficial observer is struck with the general symmetry of the construction of inorganic chemical compounds."[4] His discovery was that the regularities in the empirical formulae of the inorganic substances could be extended to the "rational formulae" of the metallo-organic compounds: Equivalents of oxygen, sulphur or iodine in inorganic formulae AsO_3, AsO_5, AsS_2, ZnO_2, HgI_2 etc. could be replaced by methyl (C_2H_3), ethyl (C_4H_5), etc. to generate the rational formulae of metallo-organic compounds

$$As \left\{ \begin{array}{l} C_2H_3 \\ C_2H_3 \end{array} \right. ,$$

$$Hg \left\{ \begin{array}{l} C_2H_3 \\ I \end{array} \right. \quad \text{etc.}$$

It is not necessary to give an atomistic interpretation of Frankland's argument. He is using the words "atom" and "equivalent" indiscriminately. This does not imply that Frankland would reject a realistic interpretation of his formulae. His adherence to the electrochemical theory and his belief in the pre-existence of radicals in compounds are reasons to suppose that the exchange of the terms atom and equivalent is based on his acceptance of a realistic interpretation of rational formulae.

This is quite different in the case of Gerhardt, whose types can be considered as an extension of the valency concept to oxygen, sulphur, nitrogen, phosphorus etc. and whose residues show a greater variety compared with Frankland's radicals. Because he rejects a realistic interpretation of his rational formulae his implicit valency concept is of a strictly conventional nature. The same holds for Kekulé's extension of valency towards carbon via the marsh gas type. Since then valency is a set of numbers associated with the elements, which with the help of certain construction rules generates a special type of rational formulae, named structural formulae. With every existing compound one and only one structural formula is associated. The determination of the "structure" of a specific compound and — related to that — the definition of the concept "structural formula," is a quite different matter to which I will return presently.

4. E. Frankland, "On a New Series of Organic Bodies Containing Metals", *Philosophical Transactions of the Royal Society of London* 142 (1852) 417-444 in: O.Th. Benfey, *Classics in the theory of chemical combination* (New York, 1963) 104.

Wurtz's 1867 reductive definition of valency does not fit into the trend of extension and sophistication of a "conventional" concept. We have to remember that Wurtz was one of the champions of the atomic theory in France and a great teacher, convinced of its expository value. Although valency could not be explained satisfactorily in the mechanical theory, Wurtz's expectations are still high in 1868. His historical introduction to the *Dictionnaire de Chimie* closes with the remark

> " ... affinity and atomicity are the two manifestations of the force which resides in the atoms, and the atomic theory is today the common basis of all our theories ... It lends a catching simplicity to laws about the composition of substances, provides insight into their internal structure, it intervenes in the interpretation of their properties, their reactions, their metamorphoses, it will later undoubtedly lend support to molecular mechanics."[5]

We arrive now at the centrepiece of Professor Nye's paper: how are the formulae and reaction equations, used in the nineteenth century, seen by contemporary chemists? We have to be careful: don't take the words of those chemists themselves too literally! When Lavoisier uses his occasional formalism for the solution of iron (♂) in nitric acid = 'nitric air' ($\triangle+$) + oxygen ($\oplus$) and water (∇): $(a\text{♂}) + (2ab\,\nabla + \frac{ab}{q}) + \frac{ab}{s}\oplus + \frac{ab}{t}\triangle+)$

$$(a\text{♂}) + \frac{a}{p}\oplus) + (2ab\,\nabla + \frac{ab}{q}\nabla) + (\frac{ab}{s}\oplus - \frac{a}{p}\oplus + \frac{ab}{t}\triangle+),$$

this could be considered a reaction equation after the insertion of the equality sign between the two sums. And although Lavoisier begs us not to "consider these formulas .. as more than simple annotations, of which the object is to ease the labors of the mind", it is clear that this formalism is much more than a heuristic instrument or a shortcut for thinking. As Gillispie remarked, Lavoisier is too modest in this very rare case![6]

If we cannot rely on the contemporary chemists themselves, we are forced to trust our own analytical powers, which might be strengthened with the help of another discipline: the philosophy of science.

Chemical formulae, I think, inform us about chemical reactions, within a certain theoretical framework. The empirical formula shows which elements and how much can be isolated from the substance identified through the formula. The theoretical frame-work is that of Lavoisier's chemistry. Rational formulae give extra information about reactions,

5. A. Wurtz, *Dictionnaire de Chimie pure et appliquée* (Paris, 1874) LXXXVI.

6. C.C. Gillispie, *The Edge of Objectivity* (Princeton, 1970) 242-245, on A.L. Lavoisier, "Considérations générales sur la dissolution des Métaux dans les acides" (1782) in: *Oeuvres de Lavoisier* (Paris, 1862) II, 509-527.

again on the basis of a specified theory. Berzelius' electro-chemical theory is an example. It would be worthwhile to isolate the formal aspects of Berzelius' chemistry. The formal system consists of a (classified) catalogue of radicals and elements, rules for the construction of acceptable rational formulae from these radicals and elements and transformation rules generating chemical reaction equations from the rational formulae. The system is based on the analysis of a restricted set of known chemical reactions and the electro-chemical theory. This theory is developed in the realm of inorganic chemistry and considered applicable to organic compounds and reactions as well. The history of the radical theory can be seen as the continuous adaptation of the catalogue of radicals and the set of construction and transformation rules, in order to be able to describe the growing network of known compounds and reactions. When transformation rules had to be accepted which were at variance with the electro-chemical theory, the whole program came under attack.

Laurent's nucleus theory did away with radicals and with the electro-chemical theory that justified the rational formulae and the transformation rules of the radical theory. The formal system of the nucleus theory generates rational formulae from a number of fundamental nuclei, C_hH_m through replacement of Hydrogen equivalents by specified groups, or through addition of members of an other group. Transformation rules generate from specified classes of rational formulae other classes. The chosen formal system is justified by the totality of the known chemical compounds and reactions it describes, not by constitutional considerations. Laurent's well-known geometric representation of the nuclei are meant to be illustrations showing hypothetical possibilities for the constitution of organic compounds, but are not justified at all by the success of the nucleus theory in covering chemical reactions.

Gerhardt's new theory of types adds the chemical classification of types to the genetical classification of nuclei and derived nuclei. Nuclei and their derivatives by addition reflect the practical possibility to generate compounds of the same class by chemical reactions from each other. The chemical types reflect analogies in reaction pattern. The type is defined through a set of typical reactions, i.e. by a specified set of transformation rules. A known compound belongs to a certain type if its reactions conform to these rules. These reactions may exhibit different typical reaction patterns at the same time; consequently, the compound will be described by different typical formulae simultaneously. Gerhardt's motto "autant de réactions, autant de formules rationelles" cannot be taken literally: it would make his formulae non-informative.

Kekulé's theory of structure can be considered a further sophistication of the nucleus- and type-theories. The formal system that can be separated from Kekulé's theory consists of a set of precise construction rules for acceptable rational formulae starting from the symbols for the elements and associated valency-numbers, again combined with a set of transformation rules for rational formulae generating reaction equations.[7] Among the construction rules are recursive rules allowing the construction of $C_nH_{m-1}CH_3$ from C_hH_m and leading for instance to constructs like

$$C\left\{\begin{array}{l} H_2 \\ CH_2(CH_3) \\ CH_3 \end{array}\right.$$

starting from $CH_{\{s\}4}$. The chemical problem, or the problem of the chemist using the theory of structure, is to find the unique structural formula of an isolated compound or to synthesize a compound of known structural formula from other compounds. In the nineteenth century the determination of structure is performed exclusively through the analysis of chemical reactions. A synthesis is performed by application of the transformation rules generating the structure aimed at.

The historical problem is to understand how structure theory, so extremely successful in its eventual form, developed out of the successes and failures of the past. Although the actors in the historical play are no logicians and at best amateur methodologists, a logical and methodological analysis of the great theories of nineteenth-century organic chemistry will certainly shed new light on the historical process, directing our attention towards fundamental differences and similarities.

One of the most conspicuous similarities of the nucleus, type and structure theories is their complete independence of constitutional hypotheses. One might say that the nucleus, type and structure theories are conventional in principle. No wonder that Laurent, Gerhardt and Kekulé considered them conventional. The interesting historical question is whether they became conventionalists because the fallacies of the radical theory drove them towards a conventional position or whether they developed their theories under the influence of an empiristic, instrumentalistic or conventional philosophical climate. Professor Nye

7. J.J. Mulckhuyse, *Molecules and Models. Investigations on the axiomatization of structure theory in chemistry* (Thesis Univ. of Amsterdam, 1960), offers a formalization without considering transformation rules, which connect the formulae with reaction equations. It will be clear that such a formal system is semantically deficient.

does not pose this question and consequently it is not my task to answer it.

Professor Nye characterized the development of rational formulae as starting from pictorial symbols and developing via linear formulae not meant to be algebraic towards non-linear constructs not meant to be geometrical. My comments have the intention to shift your attention from the representation aspects of the formulae towards their formal properties and their meaning. Not that representation is historically unimportant: effective representation makes the formal operation of the theory perspicuous and easy to handle. This ensures the impact of the theory on students and on chemical researchers, who are rarely logicians. The most interesting question of Nye's paper is how the chemists of the 19th century understood these representations. I have some reservations in following her conclusion. Most chemists — I think — understood them as reflections of a reality of atoms and molecules. Laurent, Gerhardt and Kekulé considered their rational formulae to be convenient formal symbols, which, within the formal theories in which they function, lead to equations that reflect the perceived facts of the network of known compounds and known chemical reactions. Kolbe on the other hand, considered his formulae as the reflection of molecular reality and the typical, structural and stereochemical formulae as products of an imagination run wild. This brings us at last to the tetrahedric carbon atom. Van 't Hoff's remark "The representations themselves are after all something doubtful, as is the tetrahedron itself," seems to enlist him in the conventionalist camp. Urbain's comment: "the (double) tetrahedron is a pure geometrical symbol ... a synoptic sign of writing ...," is completely beside the mark. It ignores the convenience of the tetrahedron in organizing an important domain of experimental data. Enlisting Van 't Hoff as a pure conventionalist would make it the more amazing that afterwards his stereochemical formulae appeared to be a true reflection of the geometry of molecules. Historically there is a direct relationship between early stereochemistry and constitutional considerations. The tetrahedric carbon atom did not only explain chemical reactions but also symmetries in crystals and optical rotation. These properties already induced Pasteur to the postulation of tetrahedral arrangements of constituting atoms. Laurent would have hailed these relationships as the first real connection between the theory of reactions and the theory of constitution. Van 't Hoff's doubts about the meaning of the tetrahedron reflect — I think — only the influence of the empiristic and conventionalistic tradition in the third quarter of the nineteenth century.

I suggested in my comments that it will be fruitful for historians to use recent methods of logical, methodological and philosophical analysis.

The philosophers of sciences show a growing interest in the history of science in their analysis of what science is and how it develops. The historian certainly is interested in the same questions. In his reconstruction of the past he cannot evade from these problems. The philosophy of chemistry is an underdeveloped subject. I think historians can contribute considerably to its development by the methodological analysis of the great theoretical systems of the past. I have tried to suggest that such an analysis will be fruitful for the history of chemistry as well. Presumably it could also help in understanding the complexities of the discipline of chemistry at the present.

University of Nijmegen

R.C. Olby

REDISCOVERY AS AN HISTORICAL CONCEPT

Introduction

In the last twenty-five years the history and philosophy of science has gained in rigour, sophistication, and breadth of vision. The profession has won some respect from the scientific community, but its own practitioners have not achieved a consensus as to how their subject should be studied, or even what constitute the legitimate objects of their investigations. When we carry out an intellectual stock-taking, when we look for trends, what do we find? Have all historians of science become social relativists? Are scientists for us just another primitive society with their own "folk reasoning", like the Azandi Indians about whom we have learnt so much recently?[1] Are we all talking about social legitimation, negotiated knowledge, laboratory life[2] and myths of origins?

In this paper an attempt is made to explore some of these recent trends by examining the concept of rediscovery and the example of Mendelian heredity. The resulting analysis has concentrated upon the interpretation of this example by Augustin Brannigan.[3] His sociological conception of the nature of discovery will be described and discussed.

The Concept of Rediscovery

In the last two decades the history of science has been attacked from two directions: Structuralists have claimed that we can never know what were the intentions and the motives of scientists in the past. Nor are we in a

1. P. Winch, "Understanding a Primitive Society", *American Philosophical Quarterly* 1 (1964) 307-324. M. Douglas, *Purity and Danger: an Analysis of the Concepts of Pollution and Taboo* (London, 1966) 1, 76, 86 ff.
2. B. Latour, & S. Woolgar, *Laboratory Life. The Social Construction of Scientific Facts* (London, 1979).
3. A. Brannigan, *The Social Basis of Scientific Discoveries* (Cambridge, 1981), (referred to in future as "Brannigan").

position to reconstruct their thoughts. Nor can we justify the construction of the causal relations between the science of one period and that of its successor.[4] Instead we should concentrate our efforts upon revealing the structure of the period by the study of its literary productions over a broad field, thus to expose the characteristic and pervasive features of the discourse in which the scientific writings are embedded.

Sociologists have in their turn attacked the mentalist, and intellectualist character of much history of science. They have spread scepticism as to the possibility of objective knowledge, urging instead the part played by sociological features of the scientific community-funding, career roles, political influence, social needs. At first the standard examples displaying the influence of sociological factors were dismissed as pathological, not typical science — Nazi eugenics, Lysenkoist biology, Victorian cranioscopy, etc. Recently the sociology of knowledge approach has moved far beyond that position and now claims that all science is amenable to sociological analysis because it is all socially determined. Bloor has applied the sociology of knowledge approach to mathematics, MacKenzie to statistics,[5] and Shapin to Newtonian philosophy.[6]

How have these developments affected our conception of the nature of discovery and rediscovery? First, historians have already become disinclined to concentrate so much attention upon "Great Discoveries". Philosophers have been willing to surrender discovery to the psychologist and to give their attention to the philosophical status of justification. Thomas Kuhn while denying that discovery and justification can be clearly separated, turned to Gestalt Psychology for analogies to describe how great innovations are made.[8] The sociologist Brannigan attacked mentalistic, psychological models of discovery in his book *The Social Basis of Scientific Discovery* (1981). After pointing out the inadequacies of such accounts he concluded:

> "... the societal context, not the psychology of learning, determines the discovery as a social fact. This is what the psychology, history and philosophy of science overlook."[9]

Accordingly he has developed an "attributional" theory of discovery

4. M. Foucault, *The Archeology of Knowledge* (London, 1972) chap. v.
5. D. Bloor, "Polyhedra and the Abominations of Leviticus", *British Journal for the History of Science* 11 (1978) 245-272.
6. D. MacKenzie, *Statistics in Britain, 1865-1930: The Social Construction of Scientific Knowledge* (Edinburgh, 1981).
7. S. Shapin, "Of Gods and Kings: natural philosophy and politics in the Leibniz-Clarke dispute", *Isis* 72 (1981) 187-215.
8. T.S. Kuhn, *The Structure of Scientific Revolution*, 2nd ed. (Chicago, 1970) 112 ff.
9. Brannigan, 44.

according to which the status of discovery is attributed by the Scientific Community to a given piece of work. The task of the sociologist of science is then to describe the process of granting discovery status. Let us give up searching for the psychological processes leading to the discovery being made, and instead find out by what social processes discovery status is awarded to a given piece of work.

In the case of Mendel's work the reader will recall that according to the popular view his results published in 1866 were ignored for 34 years until three botanists, Hugo de Vries, Carl Correns, and Erich Tschermak, independently rediscovered his laws in 1899 and on searching the literature they discovered that Mendel had anticipated them.[10]

Adopting Brannigan's attributional approach we would find that Mendel's work was in fact known in its time, but it was treated as a contribution to the normal science of that period. Only in a different context and 3 decades later was it seen as a major novelty deserving the status of discovery. In 1900 it could receive such a status because it could be fitted into the new perspective of the discontinuity of variation which Bateson and De Vries recognised as nature's mechanism for preventing the dilution and loss of the variability necessary for evolution to occur.

Thus it is not that Mendel discovered his famous Mendelian ratios in the 1860s, and that they were justified (repeated and confirmed) in 1900, but that Mendel completed an interesting set of experiments on the question debated at the time as to the role of hybridisation in producing new species. In 1900 the theoretical and experimental results of Mendel's work could be adapted to solve other important questions, and in this context were judged to merit the status of a discovery. So we do not need to ask questions about how Mendel viewed his own work. We need not bother ourselves with unsubstantiated claims such as that Mendel once remarked: "My time will come!"[11] Nor do we need to go on searching the literature for references to Mendel in order to confirm the assumption that his work was neglected because of a failure in the communication network of nineteenth-century biology. The work itself did not carry the status of discovery — that status had to be attributed to it by virtue of a social process.

Now by rediscovery we mean that a novel observation, generalisation or theory was made by one person and was later made by another person in ignorance of the previous person's work. It is then discovered that the observation, law or theory had already been stated before. Examples are

10. H.F. Roberts, *Plant Hybridisation Before Mendel* (Princeton, 1929, reprinted 1965), chap. xi.

11. H. Iltis, *Life of Mendel* (London, 1932, reprinted 1966), 282.

said to abound from the discovery of the rare gases, the kinetic theory of gases, the observation of sun spots, and of the planet Neptune, to the principle of natural selection.[12] These examples have traditionally been treated as unproblematic. That is to say, if the first discovery remained obscure or unknown then it was due to a failure in communication. If it was rejected — for instance when refused by a journal — Waterston's enunciation of the kinetic theory of gases turned down by the *Proceedings of the Royal Society*[13] — then it was because the discovery could not be fitted logically into the structure of then-current theory. When society reaches the appropriate level of knowledge and has reached a state of concern and awareness of problems relating to the discovery (e.g. concern over public health and sanitation and the emergence of the germ theory of disease) then the discovery will be recognised. Equally, if the discovery has been lost from view completely (e.g. Mendel's work) then it will be rediscovered in the course of the logical development of science.

Behind this view, which we will call the naive view of rediscovery, lies the assumption that facts are facts — they are unambiguous statements about a *constant* world, which exists in the form it does independent of our intellectual operations upon it. The meaning of facts is determined by the object to which they refer, not by the preconceptions and assumptions of the observer. Therefore it was possible for Scheele, Priestley and Lavoisier all to discover the same gas, although only Lavoisier called it oxygen. Turning to geography, America, like oxygen, has physical constancy — and exists independent of any explorer happening to land on it. One might say America was discovered by the Neolithic Siberians, and rediscovered by the pre-Christian Phoenicians, then by Irish monks, the Norsemen and Columbus.

The example of America, so well utilised by Brannigan,[14] is particularly instructive. If we define the discovery of America as the first report of the landing on a coast which we (with the advantage of modern knowledge) can identify as part of the American continent, then it is difficult to decide who was first. Having decided this, let us say in favour of the

12. R.K.S. Merton, *The Sociology of Science, Theoretical and Empirical Investigations*, edited by N.W. Storer (Chicago, 1973) chap. xvi.

13. S.G. Brush, "The development of the kinetic theory of gases II. Waterston", *Annals of Science* 13 (1957) 273-282.

Waterston's 1845 paper was published by Lord Rayleigh in 1893: "On the physics of Media that are Composed of Free and Perfectly Elastic Molecules in a State of Motion", *Philosophical Transactions of the Royal Society* 183A (1893) 1-79. This includes the Abstract which the Society published in 1846 — see: *Proceedings of the Royal Society* 5 (1846) 604.

14. Brannigan, chap. vii.

Norseman, Bjarni Herjolfsson in A.D. 986, followed up by Leif Eriksson in A.D. 1002, then Columbus in 1492 and Amerigo Vespucci in 1502 were rediscoverers.

On the other hand, if we define the discovery of America as the exploration of land believed to form no part of the then known continents, but of an unknown continent (a New World) not even connected to Asia, then surely we would call Amerigo Vespucci the discoverer of America as indeed the name America proclaims. Columbus clearly expected to arrive in a populated part of Asia — what the people he found there called "Guanahani" — for he took bells, buttons and beads with which to barter. On his successful return the Spanish Monarch applied to the Vatican for recognition of Spain's claim to the new lands, but viewing them as parts of Asia lying "towards the Indies".

Columbus was urged to prove that he had in fact reached the Eastern extension of the Far East by pioneering a "Western Passage". To discover such a route Columbus travelled further south, but he found the coast of Cuba in his way. At first he concluded that this coast represented the southern part of the Chinese province of Mangi. Other land, further south, must, he suggested, be a hitherto unknown extension of Asia in an easterly direction.

Vespucci, on the other hand, who like Columbus tried to find a "Western Passage" persisted to 50 degrees South without success. The enormous coastline which prevented him, he concluded, was that of a new Continent, hitherto unknown to Europeans, yet inhabited. In 1507 Martin Waldsemüller published his *Cosmographiae Introductio* showing the new lands which he called *America* as making up one quarter of the world.[15]

By our second definition Columbus had not discovered America any more than had Bjarni Herjolfsson. Rather, the recognition of the *results* of Columbus' voyages as the discovery of a new continent, by other members of society *after* Columbus' first voyage led to the identification of Columbus with the discovery of America. This, claims Brannigan, is clearly a social act.

Historians admit that discovery as a discrete act, dissolves under the microscope of empirical historical research and proves to be an artefact of faulty recollection. Shifting attention from discovery to justification has not solved the problem, for a consensus has not emerged as to the possibility of separating justification from discovery. Brannigan's answer

15. M. Hylacomylus, *Cosmographiae Introductio* ... (Saint Dié, 1507). See also: E. O'Gorman, *The Invention of America* (Bloomington, 1961).

to these disappointments is to treat discovery as a process of *attribution.* He declared:

> "In spite of the fact that the North American land masses were visited repeatedly by the fifth century Phoenicians, the Irish, the Norse as well as (by) nomadic Siberians two hundred centuries before the voyage of Columbus, this culture gives recognition for the 'real' act of discovery to Columbus."[16]

Instead, therefore, of pursuing *naturalistic* accounts (how the observation of nature leads to discovery), or *psychological* accounts (how the scientist's conscious, subconscious, and even unconscious thoughts generate discoveries), we should study the social process of granting *discovery status* to scientific work.

This approach is both illuminating and helpful. To circumvent the problems inherent in a study of a discoverer's motives and intentions, his own preconceptions of his achievements, is indeed attractive. Yet the history of science would be impoverished were we to run away from such problems. The pursuit of the personal elements in scientific creativity has already yielded far too rewarding a series of results for us to want to turn our backs on such investigations now.[17] Nor would we want to discontinue our studies of the conceptual and observational aspects of major discoveries. Consider what has been achieved by the systematic study of the Darwin and Freud archives over the past decade.[18] Why, we may well ask, should a sociological analysis of rediscovery and discovery make redundant a psychological, philosophical, or "internalist" historical analysis? Accepting that the sociological perspective has enriched and informed the history of science, and that it would represent a step backwards to reject it, our task is surely to integrate it with other perspectives. Any claim that the perspective of one discipline makes redundant those of others invites the suspicion that professional claims are being made on behalf of that discipline and at the expense of the traditional claims of philosophy, psychology, and history. Just because there is a sociology of science, a sociology of art, a sociology of politics, does *not* mean that there is no longer any psychology of these subjects.

16. Brannigan, 123.

17. H. Gruber, *Darwin on Man. A Psychological Study of Scientific Creativity* ... (London, 1974) F.L. Holmes, *Lavoisier and the Chemistry of Life. An Exploration of Scientific Creativity* (Madison, 1985).

18. M.J.S. Hodge, "Darwin as a Lifelong Generation Theorist", in D. Kohn (ed.),*The Darwinian heritage: A centennial retrospect* (Princeton, 1985) 207-243. D.Kohn, "Theories to work by: Rejected theories, reproduction and Darwin's path to natural selection", *Studies in the history of biology* 4 (1980) 67-170. F. Sulloway, *Freud, biologist of the mind* (London,1979). J.M. Masson (ed.), *The Complete Letters of Sigmund Freud to Wilhelm Fliess, 1887-1904* (Cambridge, 1985).

We are not so naive as to believe that such disciplines as sociology, psychology, philosophy, and history, are selfcontained, their territories clearly defined so that there are no mergings between their borders. Therefore let us be free to use them all as and when they offer us appropriate tools with which to investigate the history of science.

The "Rediscovery" of Mendel's Laws

The bare facts of the event are as follows: References to Mendel's work between 1866 and 1900 are few and brief. Only two of them: Focke's *Pflanzenmischlinge* (1881) and Schmalhauzen's Master's Dissertation in 1874 at St. Petersburg, actually mention Mendel's ratios. Two of them: Schmalhauzen's and Blomberg's were theses not available in the form of papers in scientific periodicals.[19]

In 1900 eight papers appeared by three botanists — De Vries, Correns, and Tschermak, all on the subject of heredity in hybrids. Each author claimed to have *discovered* the *ratios and* their *explanation* in terms of germinal segregation and random recombination. They asserted in 1900, and later, *their independence of each other and of Mendel.* Hence they have been cited *ad nauseam* as an example of *multiple* discovery. The rediscovery of Mendel's principles figures therefore in the list of sixteen multiple discoveries of A.L. Kroeber, of 1917,[20] in Ogburn and Thomas' list of 148 in 1922,[21] in Leslie White's of 1969,[22] and Merton's of 1973. Merton claimed extravagantly: "The pages of the History of Science record thousands of similar discoveries having been made by scientists working independently of one another." "The case of Mendel", Merton opined, is "too well known to need review".[23]

The key "rediscoverers'" papers are listed in Table I. Only the first, De Vries' paper in the *Comptes Rendus*, made no mention of Mendel, although, as Correns noted in 1900, De Vries used Mendel's terms *dominant* and *recessive* instead of his own terms, *active* and *latent*. None of De Vries' 1900 papers gave the numerical data for F2, only the percentages of dominant and recessive offspring, which are accounted for as the result of a process of "disjonction" between dominant and

19. For a detailed catalogue of these references see: Olby, *Origins of Mendelism*, 2nd ed.(Chicago, 1985) 219-234.

20. A.L. Kroeber, "The Superorganic", *American Anthropologist* 19 (1917) 163-213.

21. W.F. Ogburn and D. Thomas , "Are Inventions Inevitable? A Note on Social Evolution", *Political Science Quarterly* 37 (1922) 83-98.

22. L. White, *The Science of Culture*, 2nd ed.(New York, 1969).

23. Merton, *op. cit.* (ref. 12) 358.

Table 1. The four 'rediscovery' papers of 1900

Author	Title	Date sent	Meeting	Published
In the *Comptes Rendus*				
De Vries	'Sur la loi de disjonction...'	14 March	26 March	before 21 April
In the *Berichte*				
De Vries	'Das Spaltungsgesetz...'	14 March	30 March	25 April
Correns	'G. Mendel's Regel über das...'	24 April	27 April	23 May
Tschermak	'Ueber künstliche Kreuzung...'	2 June	29 June	24 July

recessive pangenes in the formation of the pollen grains and ovules. Correns and Tschermak, in their first publications, on the other hand, both independently mentioned Mendel whose paper they had discovered, each in the course of a literature search following the evaluation of his research results. It would appear that both men realized there was something significant about data which approximated to the ratios 3 : 1 and 9 : 3 : 3 : 1. In other words they had made a "discovery" of some kind, and they wanted to know whether there existed other reports already in the literature. Since these ratios were not in any obvious way connected with the aims of their experiments — the xenia question and heterosis — it is clear that they attached some importance to them, and hoped they could lay claims to originality in having *discovered* them.

In De Vries' case the evidence compiled by Kottler, Campbell, and most systematically and effectively by Meijer, strongly suggests that no such empirical discovery was made.[24] Rather, it was the receipt of a reprint of Mendel's paper by De Vries in March 1900 that, for the first time, enabled him to select from his wide and varied data those which conformed to a common feature — approximation to the ratio 3 : 1 — and also offered him a theoretical explanation in terms of germinal segregation. Hitherto De Vries had sought to explain his data in terms of the concept of the "hereditary power" or "hereditary force" of the characters crossed. This was revealed in a number of ways, only one of which concerned the proportions of individuals carrying the one or other character in the hybrid offspring. Since the hereditary strength was variable, the proportions revealed in the F2 generation could be equally

24. M. Kottler, "Hugo de Vries and the Rediscovery of Mendel's Laws", *Annals of Science* 36 (1979) 517-538. M. Campbell, "Did De Vries Discover the Law of Segregation Independently", *Annals of Science* 37 (1980) 639-655. O. Meijer, "Hugo de Vries no Mendelian?", *Annals of Science* 42 (1985) 189-232. L. Darden, "Hugo de Vries's Lecture Plates and the Discovery of Segregation", *Annals of Science* 42 (1985) 233-242.

varied, and there was no *a priori* ground for expecting particular ratios.

Whilst we can moderately confident that De Vries was dependent upon Mendel for recognition of the ratios 3 : 1 and 9 : 3 : 3 : 1 and their interpretation, the degree of dependence of Correns and Tschermak is less clear. Neither appears to have needed Mendel to tell them about the ratios, but whether Correns needed Mendel for the explanatory theory is uncertain. Since Tschermak, even after reading Mendel's paper, did not show an understanding of the theoretical explanation, he surely did not arrive at it himself.[25]

We are not, therefore, left with three *equivalent* rediscoverers. Any sense of rediscovery, on which we can speak with confidence, refers only to the ratios and not to their theoretical explanation. The dependence of these three botanists upon Mendel was much stronger than any of them were prepared to admit. In their role as "rediscoverers" perhaps their most important function was, then, to *attribute* the status of discovery to Mendel's work. But this attributional process was not merely a social process. It was an intellectual process in the sense that judgements being made in the rational evaluation of the merits of the Mendelian theory were informed by the ongoing results of the "rediscoverers'" experiments.

This process of attribution can be discerned most clearly in the writings of William Bateson. In my view he was not well prepared to receive Mendel's work. When he first read about segregation (or *disjunction*) in De Vries' *Comptus Rendus* paper he looked upon it as a *modification* of Galton's Ancestral Law of Inheritance to apply to non-blending characters.[26] When he read Mendel's paper he realized that he was not dealing merely with a modification of Galton's law. Like De Vries, Correns, and Tschermak, however, Bateson was anxious to establish the extent of application of Mendel's law. At first he treated it as the form of inheritance for non-blending characters, whilst Galton's law applied to blending characters. When he realized that the dominant/recessive relationship was not a necessary feature of Mendel's theory he opened up the possibility for applying Mendel's heredity to blending characters — as well as non-blending — treating Galtonian heredity as a special case derivable from Mendelian heredity.[27]

We can see the process of attribution at work in all the rediscovery papers, save De Vries' brief report in the *Comptes Rendus*. The early Mendelian writings of Bateson are interspersed with his evaluations of the place of Mendelian heredity in the broad subject of heredity and

25. See C. Stern in Stern and E.R. Sherwood, *The Origin of Genetics. A Mendel Sourcebook* (San Francisco & London, 1966) xi.

26. Editorial "Royal Horticultural Society: Scientific Committee", *Gardeners' Chronicle* Series III, 27 (1900), Supplement, p. 3.

27. Olby, "William Bateson's Introduction of Mendelism to England", *British Journal of the History of Science* 20 (1987) 399-420.

variation. Tschermak, Bateson, and his colleague, Biffen, all analyzed traditional concepts and laws of inheritance in the light of Mendelian theory and showed what a transformation was being effected, which in their view made traditional concepts outmoded, even erroneous.[28] It is not clear from Mendel's own work how far he believed the implications of his work extended. He believed in the "Unity of the developmental plan of organic life"[29] and he saw two basic plans — temporary association of differing elements in the hybrid leading to their segregation as in *Pisum*, and permanent association as in *Hieracium*, but he did show that in a number of genera other than *Pisum*, the segregating type of hybrid was produced.

The situation in 1900 was different. Millardet had described his "false hybrids" which fitted neither the *Hieracium* nor the *Pisum* type.[30] Correns classified hybrids into four different groups. Two of these were Mendelian in the modern sense, i.e., of the segregating *Pisum* type. De Vries held that *Pisum*-type inheritance only applied to what he called "balanced hybrids", i.e. those in which each kind of hereditary pangene is matched by alternative pangenes. The important *Progressive* variations — mutations leading to new species — were therefore not subject to *Pisum*-type inheritance since they resulted from the presence of new pangenes. Therefore De Vries did not believe that Mendel's discovery was so very important. Both Bateson and Tschermak were enthusiastic for the application of Mendelian theory to cultivated varieties of plants but doubted that it applied to inter-specific hybrids.[31]

In the period 1900 to 1909 there was an ongoing debate over the status of Mendelian heredity. What sort of a discovery had Mendel made? This debate was quite distinct from Mendel's own attribution of discovery status to his work, and the social and intellectual contexts of the two events were quite different. Hybridisation as a mechanism for species

28. E. Tschermak, "Ueber die Züchtung neuer Getreiderassen mittels künstlicher Kreuzung. Kritischhistorische Betrachtungen", *Zeitschrift für das landwirtschaftliche Versuchswesen in Oesterreich* (1901) 32. W. Bateson, "The facts of Heredity in the Light of Mendel's Discovery", in: Bateson and Saunders, *Reports to the Evolution Committee of the Royal Society* 1 (1902) 125-138; R.H. Biffen, "Mendel's Laws of Inheritance and Wheat Breeding", *Journal of Agricultural Science* 1 (1905) 4-48. On p. 7 Biffen stated: "One saw still further that many of our current theories of heredity had no real foundations, and that at the first critical test they must fail."

29. G. Mendel, "Experiments in Plant Hybridisation", transl. in W. Bateson, *Mendel's Principles of Heredity* (Cambridge, 1909) 357.

30. A. Millardet, "Note sur l'hybridation sans croisement ou fausse hybridation", *Mémoires de la Société des sciences physiques et naturelles de Bordeaux,* 4th series, 4 (1894).

31. For fuller details on these claims see: Meijer, *op. cit.* (note 24) and Olby, *op. cit.* (note 19).

Table II. Approximations to the 3 : 1 Ratio which can be found in the quantitative data of non-Mendelian literature

YEAR	AUTHOR	GENUS	PREPOTENT	LATENT	RATIO
1849	Gaertner	*Zea*	224	64	3.5 : 1
			104	39	2.6 : 1
1868	Darwin	*Antirrhinum*	88 (2)	37	2.38 : 1
1894	De Vries	*Lychnis*	392	144	2.72 : 1
1895		*Papaver*	158	43	3.67 : 1
1896			1095	358	3.06 : 1
1899		*Zea*	3161	1082 *	2.92 : 1
1899	Correns	*Pisum*	1394	453 **	3.07 : 1
1899	Tschermak	*Pisum*	1854	660 ***	2.8 : 1
			884	288	3.07 : 1

* De Vries wrote:

"about 1/4 was sugary, the 3 other quarters were starchy. The former had returned to the character of the grandmother, the latter showed that of the father and grandfather."

Dec 4 1899

** Correns recalled:

"... the operation of a principle was found in maize and peas. I did not come at first upon the explanation of the conformity to law ... it came to me at once 'like a flash', as I lay toward morning in bed."

January 23 1925

*** Tschermak recalled:

"In counting out the seed-characters the ever-recurring number relationship 3:1 could naturally not escape me ..."

January 7 1925

multiplication was not a major issue in 1900 (despite the views of Lotsy), whereas Mendel's work was chiefly devoted to the mechanism by which such multiplication might occur. The attributional process served to adapt the Mendelian theory to the new context. It also played a psychological role, as Meijer has suggested, for by appealing to Mendel's priority, Correns was able to forestall any attempt by De Vries to claim priority.[32]

Twenty years ago it was still common to use great discoveries as the focus for the exploration of the science which preceded them.The figures in this period were treated prominently if they fitted the category of precursor or predecessor — e.g. Lamarck, Diderot, and Buffon as precursors of Darwin etc. T.A. Knight, Goss & Seton, Vilmorin, and Charles Naudin as precursors of Mendel. Now rediscovery was the mirror image of this fashion for precursors. Both were built on the assumptions of the *object constancy* of the items in the external world, and the *theory-*

32. Meijer, *op. cit.* (note 24).

free characters of the empirical data of the sciences. Since T.A. Knight observed characters in *Pisum* which Mendel identified as recessive and dominant and which Correns, and Tschermak 34 years later showed to behave as dominants and recessives — they all discovered the same facts. Closer analysis shows this to be false. Equally, it has been assumed that because Mendel and Correns described the character-pairs, and postulated their separation in germ cell formation, they both arrived at the same theory. In fact, as the late Professor Heimans was the first to establish, Mendel did not invent or discover the allele or gene pair. He never inferred from the character-pair the existence of a pair of determining *Anlagen*.[33] Correns did just this and explicitly admitted that in so doing he went beyond Mendel.

Conclusion

The term "rediscovery" does less than justice to the transformation undergone by Mendelian theory in its rebirth in 1900. It obscures the contrast of context in which Mendel in 1865 and his "rediscoverers" in 1900 presented their work. It places unjustified emphasis upon the mere empirical data of the proportions of F2 progeny. It implies the theory-free nature of scientific "facts". The attributional theory of discovery enriches our conception of discovery; the "rediscovery" of Mendel's work can be understood in terms of social attribution. However, this does not render the psychological conception of discovery redundant. Moreover, the process of attribution itself involves an intellectual debate over the interpretation of experimental results and the development of theoretical concepts to account for a widening range of data. The processes of justification and attribution form two aspects, one scientific, the other sociological, of the same activity within the scientific community. Where does this leave the notion of rediscovery? De Vries, Correns, and Tschermak appear to have rediscovered the ratios in their data which Mendel found in his long before. This "minimal" sense of rediscovery tells us so little about the history of the establishment of Mendelian genetics that we should cease to organize our histories of genetics around it.

University of Leeds

33. J. Heimans, "Mendel's ideas on the nature of hereditary characters. The explanation of fragmentary records of Mendel's hybridizing experiments", *Folia Mendeliana* 6 (1971) 91-98. See also O. Meijer (ed.), "J. Heimans (1889-1978). Mendel on violet spotting in his *Pisum* note", *Proceedings of the Koninklijke Nederlandse Akademie van Wetenschappen* 89 (1986) 203-216. See also: R. Olby, "Mendel no Mendelian?", *History of Science* 17 (1979) 6372. Reprinted with minor revisions in Olby, *op. cit.* (note 19) 234-258.

Lewis Pyenson

PURE LEARNING AND POLITICAL ECONOMY: SCIENCE AND EUROPEAN EXPANSION IN THE AGE OF IMPERIALISM*

1. *Introduction*

In his Massey Lectures for 1974, sponsored by the Canadian Broadcasting Company, George Steiner has recalled Claude Lévi-Strauss's discussion, a generation earlier, of what happened when men and women from the North Atlantic world discovered peoples existing in innocence and unity with nature:

> "Coming upon these shadows of the remnants of Eden, Western man set out to destroy them. He slaughtered countless guiltless peoples. He clawed down the forests, he charred the savannah. Then his fury of waste turned on the animal species. One after another of these was hounded into extinction or into the factitious survival of the zoo. This devastation was often deliberate: it resulted directly from military conquest, from economic exploitation, from the imposition of uniform technologies on native life-styles. Millions perished or lost their ethnic heritage and identity."

It made little difference if the invaders were mercenary or spiritual, if they sought to rob or to instruct:

> "The gifts which the white man had brought — medical gifts, material, institutional — proved fatal to their recipients. Whether he came to conquer or to proselytize, to exploit or to medicate, Western man brought devastation. Possessed, as it were, by some archetypal rage at his own exclusion from the Garden of Paradise, by some torturing remembrance of that disgrace, we have scoured the earth for vestiges of Eden and laid them waste wherever we have found them."[1]

These were, in Steiner's view, among the principal discontents of the civilizing mission.

*Research has been supported by the American Council of Learned Societies and the American Philosophical Society. I continue to be indebted to the staff of the Interlibrary Loan Service at Vanier Library, Concordia University.

1. Steiner, *Nostalgia for the Absolute: Massey Lectures. Fourteenth Series* (Toronto, 1974) 32.

However one views the mission — as an impetus to or as a justification for expansion — it will be evident that the final product has generally been a world away from the intended result. Despite nearly a century of resident French law professors and magistrates, Napoleonic code lies shredded in present-day Iran and Vietnam. The tradition of political diversity and freedom of expression, so central to the Netherlands, is conspicuously absent in modern Indonesia. Manhood suffrage, a tenet of twentieth-century Britain, is nowhere to be found in Pretoria's South Africa.

The reflection of European canons is frequently distorted in former colonies and satrapies. Not so when dealing with force and those who control it. Generalissimos and field marshalls the world over wear the jackets and caps of British junior officers during the First World War, albeit weighted down with a bit more wampum. Roughnecks, game wardens, and nurses also show a confluence of couture. Instruments of record are standard issue: In place of clay tablets, scrolls, and tree-bark one finds *books* and the side arm of educated women and men — the ballpoint pen. The engines of power — machine guns, automobiles, hydroelectrical power stations, nuclear reactors — all look basically alike. Here small variations on common themes reveal much about cultural values and a civilization's heritage. The presence or absence of a Western cravat in a military uniform, of an abacus in a market town, of a crucifix in a hospital ward, these give voice to the varieties of cultural adaptation.

While costume, architecture, modes of conveyance, means of destruction, and even the design of machinery easily admit of some variation, the same may not be said about the mathematical laws ascribed to the physical world. One finds $F=ma$ and $PV=nRT$, not $F=ma^2$ or $PV=nRe^T$, whether taught in Rabat or Bandung, Tananarive or Quito. Different from the case of the qualitative sciences, which provide evidence of astonishing cross-cultural mutations, on the periphery of the Western world physical laws are often conveyed unsceptically to a fault, as if local

2. I do not have a sense for the geographical distribution of heretical science, but no culture has a franchise on unfounded speculation or implausible deduction. Three cases from the Southern Hemisphere are instructive in this regard: Argentina's Ronald Richter, as portrayed by M. Mariscotti in *El Secreto atómico de Huemul: Crónica del origen de la energía atómica en la Argentina* (Buenos Aires, 1985); Brazil's César Lattes, as interviewed by J.Hamilton Ribeiro, "César Lattes, o Quixote da cîencia", *Século 21*, no. 3 (1980) 48-55; and New Zealand's A.W. Bickerton, as considered by R.M. Burdon, *Scholar-Errant: A Biography of Professor A.W. Bickerton* (Christchurch, 1956).

Immanuel Velikovskys and Herbert Dingles and David Bohms and Joseph Webers have been reluctant to publicize unfashionable scientific points of view.[2] It would be a mistake to ascribe such uniformity to practical concerns: Newton's laws are no more required for building reliable bridges than Copernicus's system was an indispensable element of sure navigation. Manipulating the mathematical laws of nature is seen as an unambiguous sign of having become civilized, in much the way that European capitalists patronized nineteenth-century science in order to appear before the world as cultured men. This observation is central to what follows: Because *discourse* in the exact sciences is insular, *activity* in physics, astronomy, and mathematics is especially well suited as a cultural probe.

Beyond the North Atlantic world, research and teaching in the exact sciences have for the past 150 years closely followed the pattern set in the principal industrial nations of Europe and North America — with Western captains being variously forced on and invited into non-Western settings. Studying the diffusion of these sciences bears especially upon our understanding of cultural imperialism. Metropolitan strategies for spreading science in a 'national' mold and peripheral responses to imperialist incursion both come into clear focus when restricted to the activity of physicists and astronomers as vectors of transmission. Practical considerations, such as finding cures for tropical diseases or deposits of gold and oil, recede in favor of examining how largely impractical scientific discourse served the ends of empire. Ultimately, such an inquiry extends the promise of deciding the extent to which 'pure' scientific discourse has, in the recent past, been insulated from socio-economic reality.[3]

The following pages expand on one feature of cultural imperialism and exact sciences, that of imperialist strategies for creating institutions abroad. The discussion follows the trajectories of scientists acting as vectors of diffusion, but it does not presuppose an identity between individual career aims and national policies or goals. Western scientists on the periphery will be seen to have worked in patterns reflecting their metropolitan training. In seeking to explicate the laws of the universe, they found inspiration from the work of their own teachers. At the same time, no different from physicists or astronomers situated in a provincial

3. These points have been elaborated in my book, *Cultural Imperialism and Exact Sciences: German Expansion Overseas, 1900-1930* (New York and Berne, 1985).

European setting, they generally adapted their styles of research to local conditions. They were, at all events, caught in a system about which they were poorly informed. Referring to the ferocious exploitation of Latin America by the North Atlantic world, Eduardo Galeano has observed:

> "For its foreign masters and for our commission-agent bourgeoisie, who have sold their souls to the devil at a price that would have shamed Faust, the system is perfectly rational; but for no one else, since the more it develops, the greater its disequilibrium, its tensions, and its contradictions."[4]

Given their self-image as representatives of higher learning, the physicists and astronomers *in partibus* kept at arm's length from forced-labor gangs and plantation squalor. They also fulfilled a role as agents of imperial design, but not without experiencing something of the uncertainties faced by Galeano's wretched victims. The examples to come bear on the dialectic of master and servant, and the counterpoint of tradition and innovation.

Comparative history will cast strategies of cultural imperialism in bold relief. The technique of comparison, wielded incisively and convincingly by Marc Bloch and his followers, requires most centrally that one phenomenon be considered in several settings. Blochian comparatists would reject, for example, an attempted comparison between German professors of paleontology and German professors of physics at the National University of La Plata, Argentina; nor would they admit to the comparative corpus a study of French, Italian, American, and German astronomers at that university. In the present text, two levels of comparison are at work on the phenomenon of the overseas diffusion of the exact sciences. Interest is focused on comparable settings that received institutions inspired by different imperialist nations. It will be seen that France, Germany, and the Netherlands each implemented a special vision of how best to secure the implantation of their wisdom and cultural institutions.[5]

The settings under consideration naturally divide into four groups. First come French and German incursions into early twentieth-century China.

4. E. Galeano, *Open Veins of Latin America: Five Centuries of the Pillage of a Continent*, trans. C. Belfrage (New York, 1973) 14.

5. The issue of comparative history is considered in Pyenson, *Cultural Imperialism* 295-9, and in L. Pyenson and S. Sheets-Pyenson, "Comparative History of Science in an American Perspective", *History of Science in America: News and Views* 3, no. 3 (1985) 3-7. Current debate on the status of comparative history may be found in the pages of the *American Historical Review* 85 (1980) 753-857, 1055-66; 87 (1982) 123-43; and also in *Comparative Studies in Society and History* 22 (1980) 143-221.

Second are French and Dutch initiatives in Southeast Asia, in Indochina and Indonesia. Then come French and German astronomers and physicists in South America — Ecuador and Chile. Finally are found French and Dutch sorties into the British Empire at locations that might have been sympathetic to non-English speakers: French Canada and Afrikaans South Africa. The institutions of higher learning set down in these places have been situated beyond the pale for academic historians of science, men and women who have over the past generation concentrated on documenting the momentous rise of knowledge in Europe and the United States. As our century draws to a close, however, I believe that the forgotten universities and observatories will loom large, much as historians have come to focus attention on peoples and civilizations bypassed by an army of writers afflicted with unwavering, Eurocentric vision. Understanding science that until now has remained largely without a history will be essential for arriving at a fuller appreciation of modern research as an inter-cultural activity.

2. *Imperialist Shanghai*

Three institutes of higher learning in the imperialist enclaves of greater Shanghai provide a natural point of departure for distinguishing French and German strategies of cultural imperialism in the exact sciences. Each of the three was established to prepare Chinese students for careers in commerce, engineering, and public health, and so in them pure learning arrived as a fillip. Each institution, while ultimately seeking to provide instruction in Chinese, had to protect its metropolitan language against the clear perception that mastery of English paved the road to material success.[6] Each had to contend with uncertainties regarding the annual financial contribution provided by the Chinese federal authorities.

The first of the three to be established was l'Université l'Aurore, founded in 1903 by the Jesuit mission at Shanghai. By the outbreak of European hostilities in 1914, the Aurora had three sections offering instruction in French: letters and law, mathematics and civil engineering, and medicine. It sprawled over a campus of seventeen acres in the Shanghai suburb of

6. The Ministry of Foreign Affairs in Paris emphasized to the French ambassador in Peking that French educational plans had to take account of the fact that in China "the language of commerce is English." Paris, Ministère des Affaires étrangères [henceforth MAE]. Série Asie 1918-1940, dossier Chine vol. 485: Ministry of Foreign Affairs to A. de Fleuriau, 22 Aug 1922.

Zikawei. A score of professors taught around 200 students. Physics courses began in 1911, taught in French by a graduate of the Paris Ecole supérieure d'Electricité, Louis de Jenlis. Separate facilities for the exact sciences came about around 1920, when the French Ministry of External Affairs granted 30,000 francs for laboratories, 90,000 francs to endow three chairs, and 450,000 francs for capital developments. In this way, the Quai d'Orsay effectively purchased control of the university. It specified that the Aurora would have to be directed by a French citizen, that its professors would have to be mostly French, that the lay professors had to be graduates of French universities or *grandes écoles*, and that all these appointments existed at the pleasure of the civil authorities in Paris. Professors under the new regime, like those under the old one, expressed no interest in research. Graduates of the exact — science faculty found employment as electrical or mining engineers: One came to direct the magnetical service at the Tsingtau observatory, an institution founded by the Germans and then mismanaged by the Japanese conquerors. The Aurora, conceived to propagate the faith in the French language, came to be a pearl on France's imperial string of privately-managed but centrally-funded finishing schools.[7]

The second Shanghai institution was the Tung-Chi University, a creation of German commercial and government interests in the period before 1914. Like the Jesuit Aurora University, the German institution consisted of engineering and medical faculties. Like the Jesuit school, the Tung-Chi was privately endowed, receiving between 1906 and 1914 well over 500,000 marks. And, like the Aurora, the Tung-Chi sought to provide instruction ranging from trade-school exercises to university courses. The quality and level of education at the German university, however, left the French university far behind. The Tung-Chi's medical faculty offered doctorates, and the engineering faculty drew favorable comment from nonpartisan observers. It continued to flourish after 1914. Immediately following the time that a group claiming to be the federal government of China broke diplomatic relations with Germany, French gendarmes seized the German university, despite the fact that it was in law a Chinese institution. Although the French removed many of the furnishings, they were unable to shut down the associated teaching hospital. After 1919,

7. The Aurora is considered in Pyenson, *Cultural Imperialism and Exact Sciences* 273-4, and in L. Pyenson, "Pure Research, Jesuit Institutions, and Metropolitan Political Ambitions: The Evolution of French Policy Overseas", *Actes du IVe Colloque, L'Influence en Chine de doctrines venues de l'Occident* [Chantilly, France, 8-11 September 1983] (Taipei/Hong Kong/Paris, to appear).

the Tung-Chi relocated on a new campus in Woosung, a Shanghai suburb. Handsome donations from German industry provided for impressive buildings. Exact sciences were taught by philosopher Joseph Drexler (who provided one of the first serious treatments of general relativity in the Chinese language) and Paul Böning (who published widely on the physics of electrical discharges in liquids), the first certified with a *Dr. phil.* in philosophy of science from the University of Greifswald and the second with a *Dr. Ing.* from the Dresden Institute of Technology. The university emphasized the scientific bases of technology and medicine, seeking, according to an English observer in 1926, to "stimulate original research." It was under the administration of Chinese scientists who sported German credentials.[8]

After the French authorities appropriated the buildings and installations of the Tung-Chi University in 1917 (these were located in the French Concession), they wanted to be able to show that French *savants* could continue that which had been begun by German *Wissenschaftler*. When article 134 of the Versailles peace treaty gave France clear title to the German property, the Quai d'Orsay transformed it into, in 1921, the Institut franco-chinois d'Industrie et de Commerce. The tribulations of this third Shanghai institution under consideration, easily the least impressive of the three, indicate the limitations of French authorities who implemented policies of cultural imperialism.

Unlike their Jesuit and their German counterparts, who were flexible about beginning university-level instruction modestly and optimistic about gradually upgrading standards, French administrators who ran the Institut were obsessed with fixing for all time the school's scholarly magnitude in the star-studded educational firmament. According to the view of French policy makers, the Institut was to be a school for secondary technical instruction; it would produce "technicians, foremen, supervisors, and clerks." The director of the school, however, Charles B. Maybon, a graduate of the Paris Ecole centrale des Arts et Manufactures and holder of a *doctorat-ès-lettres*, was an old Asian hand. He sought, like his educator counterparts in Shanghai, to run a real institution of higher learning — not an Ecole d'Arts et Métiers (in the French system a *secondary* school), but a university-level school like his alma mater, the Ecole centrale. From the beginning, he set out to reorient the Institut toward training "theoretical Engineers." He recruited students who

8. On the Tung-Chi University see Pyenson, *Cultural Imperialism and Exact Sciences* 281-92.

already had their secondary-school diplomas. The Chinese educational authorities, who licensed foreign schools, decided to place the Institut in the category of Professional Schools, First Class, something short of university rank. Maybon would not have it, as the programs in schools of this category were in his view far below those of the Institut. Students at the Institut went on strike in support of Maybon's position. The French ambassador in Peking and the French consul in Shanghai would not have Maybon. He was sacked in 1924.[9]

In defending his "theoretical" direction, Maybon emphasized that he was responding to Chinese realities. One could not reflect the French system on the China coast without suffering some distortion, and in his view the way to proceed was in an optimistic vein:

> "One must absolutely not be mistaken about the value of this word [university], which does not only represent an institution of higher learning as in France, but an establishment which, founded in imitation of American or English universities or colleges, does not distinguish between secondary instruction and higher learning, [between] the rigorous limits set in France between these two orders of instruction. From the point of view of technical instruction, which is of special interest here, an establishment like a French Ecole d'Arts et Métiers, belonging to French secondary technical instruction, would be classed as an establishment of higher learning in China."[10]

These are the words of an administrator in deep trouble, someone hoping to have forgiven a transgression against metropolitan authority, knowing that absolution would not be granted. Maybon's vision was, indeed, a special one. Different from his German precursors, he could not or did not want to obtain local funding for his school, an omission for which he was rebuked by the Ministry of Foreign Affairs in Paris.[11] And in order to operate in the black, he willingly sold the fine German furnishings that his school had inherited from the previous occupants.[12]

9. The early history of the Institut may be found in various documents surrounding the student strike and Maybon's dismissal which are kept in the MAE, Asie 1918-1940, Chine 485. Report of commission of inquiry, appended to A. de Fleuriau to Ministry of Foreign Affairs, 6 Apr 1923; Maybon's reply, "Observations sur le rapport de la commision désignée pour étudier la réorganisation de l'Institut franco-chinois d'Industrie et de Commerce", n.d.; and a cabinet note of Oct 1925.

10. Maybon, "Observations sur le rapport".

11. MAE. Asie 1918-1940, Chine 485. Ministry of Foreign Affairs to Embassy in Peking, 22 Aug 1922.

12. *Ibid.* Maybon, "Institut technique franco-chinois; rapport sur la marche des études pendant le Ier semestre de l'année scolaire 1923-24". The sale, to the Société franco-chinoise de Construction métalliques et mécaniques, netted less than $51,000 (presumably a figure in Mexican silver dollars — the coin of the imperialist realm in

On the Quai d'Orsay, French mandarins made their decision by the middle of 1923: The Institut would become a trade school, Maybon would be recalled to France,[13] and the new Institut would be administered by the Chinese.[14] The long and the short of it was that the Institut had to be given "a purely technical character." Maybon "had judged exactly neither the spirit of young Chinese nor the situation of the country. His error is excusable," the French ambassador in Peking emphasized, "but his enterprise failed."[15] The decision on the part of the Ministry of Foreign Affairs was more informed by the realities of the French position in China than Maybon might have thought. In the first place, as the French ambassador to China noted in 1922, the time was not yet ripe for a major French offensive in higher education. The Americans, for example, had used much of their revenue from the Boxer indemnity (the French part of which contributed to support the Institut at Shanghai) to set up fine colleges, while the Chinese were bit by bit pushing them out the door, discouraging good directors and teachers, and keeping only those who danced to the Chinese tune. Furthermore (according to the French ambassador) a French teacher in China

> "risks becoming alienated from his country and not carrying out useful work. He will find himself in a situation that is hardly intellectual at all, very poorly educated, perfectly disorganized. Europeans will offer him few resources, for China is not, at the moment, a very pleasant place to live."

The Chinese were filled with inertia, and it was impossible to push them. "It is necessary to take them like children, like big, vain, children, who believe themselves to know everything without having learned it and who do not tolerate intellectual superiority from foreigners."[16] In the ambassador's view, France would do well to wait, let the Chinese organize things, and move in later. At the same time, the Chinese authorities in Paris sought to set up a real Ecole d'Arts et Métiers, the graduates of which could go on for further studies in France or find posts directly in industry. A representative of the Chinese Ministry of Communications specifically asked that Maybon be relieved. Maybon was "quite a pleasant man ... *but not a technician.*" His post was to go, according to the Chinese messenger, to an "eminent engineer" from

China — although it must be noted that figures in Indochinese piasters were also indicated at this time by a "dollar sign").

13. MAE. Asie 1918-1940, Chine 485. E. de Peretti de la Rocca (Paris) to A. de Fleuriau (Peking), 15 Jun 1923, telegram.

14. *Ibid.* Wilden (consul in Shanghai) to A. de Fleuriau (Peking), 28 Mar 1923.

15. *Ibid.* A. de Fleuriau (Peking) to Ministry of Foreign Affairs, 6 Apr 1923.

16. *Ibid.* A. de Fleuriau (Peking) to Marx (Affaires étrangères), 16 Aug 1922.

Paris, someone who would be able to affiliate the Institut with a metropolitan school. The proposed new direction would especially help in the "fight against competition from the English and the Germans."[17]

French contempt for Chinese intellect combined with fear of being eclipsed by American and German rivals to bring an end to Maybon's enlightened plans. The Americans, amateur high-rolling bumblers, posed no serious problem, but the Germans were to be treated with respect. Given the provenance of the buildings and appliances of the Institut franco-chinois d'Industrie et de Commerce, the Tung-Chi University loomed as a constant reminder of unrealized aspirations. The French ambassador in Peking noted in 1923 that the Tung-Chi medical faculty worked well while the engineering faculty "evoked disappointment." The ambassador continued:

> "After eighteen months, the Germans had to recognize that their combined program of theoretical instruction and laboratory exercises was not suited to the country and that the Chinese student refuses to deliver himself to the manual tasks of the worker."

They had been forced to divide the engineering faculty in half: one part for "theoretical engineers," the other for practical studies. In complete ignorance or perhaps in willful disregard of the important positions achieved by Tung-Chi graduates, the ambassador asserted that the engineering faculty had produced only two men "of quite mediocre value," one for Siemens and one for the Shanghai tramways. It was a case of sour grapes, for, as the ambassador noted farther along, graduates of the Shanghai Institut could find positions only with the French and Belgian railways.[18]

In 1924, Maybon handed over his Institut to H. Civet, a graduate of the Paris Ecole d'Arts et Métiers, but not without a parting shot at what he felt was a recipe for disaster. Casting an eye over "the three most painful years" of his life as an educator, he cited the opinion of Jesuit missionaries that "manual labor is too discredited among the educated to permit technical education, that which we call *Arts et Métiers*, to have a

17. *Ibid.* Note of a conversation with Tsang-Ou, delegate of the Ministry of Communications of the Republic of China, 2 Aug 1922, recorded by Clichant of the Service des Oeuvres.

18. *Ibid.* A. de Fleuriau to Ministry of Foreign Affairs, 6 Apr 1923. For a survey of the positions occupied by Tung-Chi graduates, see Pyenson, *Cultural Imperialism and Exact Sciences* 285.

chance of success."[19] He must have wondered what would become of the staff that he had assembled: mathematician Ho Lou, *licencié-ès-sciences mathématiques* from the University of Louvain; electrical engineer Shan Tiwan, a graduate of the Institut polytechnique of Grenoble.[20] Such pretentions as they may have had to higher learning would never be fulfilled at the Institut franco-chinois.

The decisions made in Shanghai reflect general French policy throughout China. The authorities were interested in establishing a Franco-Chinese University in Peking, insofar as it could serve as a vehicle to prepare French teachers for positions in Chinese schools and Chinese students for "more advanced study first in France and later in Peking."[21] The Quai d'Orsay authorized Boxer indemnity funds toward financing a French naval medical school in Tientsin, which had been in operation since 1901 when Peking provided a campus, but the idea was always to send the best graduates to France for a medical doctorate. The Tientsin operation survived Chinese cutbacks in the early 1920s only through the benevolence and influence of the Bank of Indochina in Peking.[22] There is no evidence to suggest that any of the French functionaries who came to China to teach exact sciences at Tientsin, Shanghai, and elsewhere had the least interest in stimulating scientific research.

3. *Southeast Asia*

Shanghai, and China in general, presented a special problem for European imperialists. Elsewhere in Asia things were more straightforward, administratively speaking. Civilizations were subjugated by force and ordered to pay tribute in labor, money, or goods. The Dutch managed to squeeze all manner of agricultural commodity and raw material out of their East Indian colony. The French exercised less finesse and enjoyed correspondingly fewer rewards from their adventure into Indochina. Strategies for entrenching French and Dutch civilizations in their respective colonies, while suggested in styles of administrative rule, are clearly revealed by the role of exact sciences at institutions that occupied homologous niches in colonial ecology, the University of Hanoi

19. *Ibid.* Maybon, "Institut technique franco-chinois ... 1923-24".
20. *Ibid.* The teaching staff are listed in a report by Maybon and Hou Wen-yao to the Consul General of Shanghai, 8 Jun 1923.
21. *Ibid.* A. de Fleuriau (Peking) to Marx, 16 Aug 1922.
22. *Ibid.* Saussine (French consul in Tientsin) to Dr. A. King (Director of the Tientsin Naval Medical School), 28 Dec 1923.

in Indochina and the Institute of Technology at Bandung, on Java. We begin with the French civilizing mission.

3a. Higher Learning in Southeast Asia: French Vietnam.

Higher education in twentieth-century French Indochina, both in its institutions and in its teaching personnel, provides a graphic illustration of the imperial policy of associationism. This notion, identified by Joseph Buttinger as "a generous form of cultural imperialism," redefined the traditional French *mission civilisatrice*. Uncivilized races (and the French continued to the last to govern Indochina as an amalgam of indigenous peoples with differing intellectual and moral proclivities) were to become brothers of the French, *français de couleur*. These races, inhabiting *la France d'outre-mer*, were not yet ready for the privileges of French citizenship. They were to be educated in colonial schools and trained for subordinate administrative positions; any native considered fit for advanced instruction would be sent to repeat his schooling in the metropolis. It was the French system of *apartheid*, but with a perverse variation. Existence in the colonies served the greater glory of the metropolis; in the colonies, one was a mere associate of French liberty and French learning.[23] In Indochina, the French set up a primitive network of inferior schools in the vernacular and in the language of the conqueror. Where literacy had been widespread and valued, both in Vietnamese and in Chinese, and where morality was clearly set by Confucian ethics, French policy produced a nation of illiterates inclined to rebel against authority.[24]

Associationism had, however, no clear end. It was not evident how to graduate from conquered to free status. One could only follow one's destiny as a subject race, eventually to produce a culture that could respond gratefully to the inventories and questionnaires of metropolitan administrators — to invent an overseas Alsace or a non-Latin Corsica. Knowledge and learning, then, had to be epiphenomenal at colonial universities. The main idea was to produce servility toward and reverence for the metropolis. As the ideals of the Third Republic were those of the First, instruction in them simply produced resentment and dissatisfaction

23. J. Buttinger, *Vietnam: A Dragon Embattled*, 2 vols. (New York, 1967) I, 75-6. The triumph of associationist ideology, which replaced the notion of assimilationism, is persuasively told by R.F. Betts, *Assimilation and Association in French Colonial Theory, 1890-1914* (New York, 1961).

24. Buttinger, *A Dragon Embattled* I, 486-7.

among the few selected to be future *collaborateurs*. The result, on the part of French educational administrators, was vacillation between merchandising the ideals of French culture and denying the existence of these ideals. Colonial education followed the whim of the governor general, and it rose or fell depending on the man who happened to be in power. L' Université Indochinoise at Hanoi was the invention of Henri Gourdon, appointed as Indochina's first director general of public instruction by Governor General Jean Baptiste Paul Beau in 1905.[25] The proclamation of the new university, in May 1906, indicates that it emerged as the pendulum of the administration swung toward a liberal maximum. The university was designed "to promote in the Far East, above all through the intermediary of the French language, knowledge of European sciences and methods." The university's governing council, presided by Gourdon, included the directors of its various constituent schools and of the other scientific institutions of the colony, as well as professors designated by the governor general to represent other services and interests. University affiliation for existing institutions would be negotiated by the council. The grand vision included schools of law and administration, sciences, medicine, civil engineering, and letters, to be located in Hanoi or Saigon. The university was open to the indigenous peoples of Indochina, who had to have completed secondary school, *as well as* to European colonials, for whom special standards of admission would be arranged. The most apt Indochinese students were eligible for scholarships to continue their studies in France.[26]

The Hanoi branch of the university opened in October 1907 with sections in sciences, French literature, and French law. To judge from the courses offered, all of which being perforce introductory ones, the level of instruction was barely that of a European secondary-school. Mathematics and mechanics, taught by a graduate of the Ecole centrale des Arts et Manufactures, went up to trigonometry and certain "industrial applications." Physics, remanded to a medical doctor in the service of the colonial army who happened to have been an assistant in the subject at the University of Bordeaux, considered electricity, heat, optics, and acoustics from the point of view of "applications." A pharmacist with the colonial army, holding a doctorate in pharmacy and having been an

25. E. Chassigneux, *L'Indochine* (Paris, 1932) 513 [*Histoire des colonies françaises*, G. Hanotaux and A. Martineau eds, vol. V].

26. Aix-en-Provence, Archives d'Outre-Mer [henceforth Aix]. Indochine 48042. Governor General, arrêté of 16 May 1906, no. 1514a.

assistant in chemistry at the University of Nancy, taught the elements of chemistry. A graduate of the Ecole centrale des Arts et Manufactures taught industrial chemistry. The chief medical doctor of the colonial army lectured on physiology. The only professor with research credentials was Philippe Eberhardt, holder of a *doctorat-ès-sciences naturelles*, who taught botany and zoology.[27]

Governor General Beau returned to France in February 1908. As wheels turned slowly to provide Indochina with a new head man, Beau's conception of the Indochinese University sustained intense criticism. Henri Gourdon tried mightily to deflect the distortions and malicious slander directed against the principal jewel in his educational diadem. In a long defense of his general educational policy, presented to interim governor Louis Alphonse Bonhoure in August 1908, he emphasized that the word "university" had given rise to much misunderstanding:

> "Words have their destiny, especially in France where one is attached, by national deficiency, more to words than to things. The foregoing [university] sustained, before the thing itself was in the process of being organized, the most acerbic criticisms."

Beau had continually tried to phrase the new university in terms of local considerations and the ends of "foreign policy," but "the name of the establishment earned him the worst enemies." Had the courses in the university carried the title,

> "which was more exact, of practical courses in science and law (there are only three courses in letters), no one would have thought to criticize them and the ridiculous legends circulating about the university would never have been born."

Gourdon listed and disposed of the most extreme slanders. He emphasized that the university cost virtually nothing, that its staff required no large payments, that the 94 original students were drawn not far and wide from the working classes but from the pool of administrative talent in the neighborhood of Hanoi, that of those 46 who survived the first semester 41 continued for the second, and that, finally, primary education had hardly been "sacrificed for extravagant, restive, and dangerous instruction."[28] The new governor general, Anthony Wladislaw Klobukowski, arrived the next month and promptly shut down the university.

27. *Ibid.* Henri Gourdon, *Université Indochinoise, Section de Hanoi, Règlement intérieur* (Hanoi, 1907).

28. Aix. Indochine 7707. Henri Gourdon to Governor General, 14 Aug 1908.

It has generally been argued that Klobukowski closed the Indochinese University because he saw it as a hotbed of sedition, fueling the uprising of 1908.[29] Klobukowski, however, was hostile in principle to advanced learning. He quickly acted to close the Permanent Scientific Mission to Indochina, established in 1902 to inventory the natural wealth of Southeast Asia. The mission, following in the tradition of nineteenth-century missions in the Mediterranean and Mexico, consisted of several biologists selected by and responsible to the Paris Academy of Sciences. Beginning in 1905 the mission cost the Indochinese colony some 20,000 piasters for five European scientists and native staff. Between 1904 and 1908, European personnel and matériel alone came to over 86,000 piasters.[30] The mission was clearly intended to undertake pure research. Edouard Alfred Marie Broni, acting governor general in Beau's absence, wrote in 1902 that the geological and paleontological section of the mission was to conduct "exploration, from the point of view of pure science," that "its role is that of an organ of pure investigation, in the order of natural sciences, completing the program of explorations to be carried out by the botanists, the zoologists, and the anthropologists." This was in contrast to the Mining Service, soon to be reconstituted, which sought "above all, immediate, direct, and utilitarian application of knowledge" acquired about the mineral riches of the soil.[31]

Eight days after arriving in Indochina, Klobukowski met with Louis Boutan, the director of the scientific mission, who promptly asked to double his budget of 29,000 piasters. When Klobukowski queried Boutan about the "practical character" of the mission, Boutan replied that practical gain would result by application of the scientific method of observation. Klobukowski then travelled to Saigon to meet with the other resident member of the mission, Armand Krempf, who asserted: "I do nothing practical, ... I do pure science." Both Boutan and Krempf emphasized that they were responsible to the Museum d'Histoire naturelle in Paris, not to the colonial regime, and that the fruit of their research would appear in metropolitan journals rather than in colonial memoranda. Klobukowski concluded that Boutan was "consolidating his

29. Buttinger, *Vietnam: A Dragon Embattled* I, 66.

30. Paris, Archives nationales [rue des Francs-bourgeois, henceforth AN]. F-17, 17261: Evaluation of the mission, 6 Jul 1908; "Dépenses de la Mission scientifique permanente de 1904 à 1908".

31. Paris, Archives nationales, Section Outre-Mer [Henceforth SOM; these archives, until 1986 located at the rue Oudinot, are scheduled to join the material already removed to the overseas archives in Aix-en-Provence.] Indochine X 50 (1). Broni to Minister of Colonies, 2 Oct 1902.

position in the learned world of the metropolis," and Krempf was holding down a lucrative post while preparing a doctoral thesis.[32] Only the third scientist of the mission, Philippe Eberhardt, escaped Klobukowski's wrath, for Eberhardt had had the good sense, after his teaching post at the Indochinese University evaporated, to be granted leave from the mission to serve as preceptor to the King of Annam.

The perpetual secretary of the Academy of Sciences, mathematician Gaston Darboux, wrote to the minister of public instruction, emphasizing that England organized expensive scientific expeditions in its colonies, "an example to be imitated from all points of view."[33] The Ministry of Public Instruction naturally protested Klobukowski's action — the activity of the scientific mission, after all, was being supported on the budget of the Ministry of Colonies — but Klobukowski would not be swayed.[34] He wrote to the minister of colonies:

> "Whatever may be the value of the arguments developed by the professors in the Faculté des Sciences of Paris, I persist in believing that in the present state of finances of the colony, it is not permitted me to allow the continuance of an institution, the scientific interest of which I do not fail to perceive, but an institution that, despite the allocation of an annual budget of nearly 100,000 francs, has not produced any practical result."[35]

The permanent scientific mission ended just as the Indochinese University had ended.

Klobukowski's successor, Albert Sarraut, slowly reversed the precipitous decline of higher education. He looked into those among the university's schools which had survived in other forms. In 1912, he received a report on the Ecole indigène des Travaux pratiques, the civil-engineering school, suggesting that "far from being abolished, this school should be reorganized on a more serious basis, in such a way as to form a nursery for technical agents" who could replace French foremen, clerks, and technicians.[36] Five years later Sarraut was once more governor general of Indochina. Replying to Sarraut's inquiries about the Aurora University, the French consul in Shanghai misleadingly identified it as the only imperialist institution of higher learning in China which instructed

32. AN. F-17, 17261. Klobukowski to Minister of Colonies, 10 Jan 1909.
33. *Ibid.*, Darboux to Minister of Public Instruction, 17 Oct 1908.
34. *Ibid.*, Minister of Colonies to Minister of Public Instruction, 15 Mar 1909.
35. *Ibid.*, Klobukowski to Minister of Colonies, 16 Apr 1909.
36. Aix. Indochine 43682. Chef, Service des Affaires politiques et indigènes to Governor General, 19 Jul 1912.

Chinese exclusively in their own language; he continued that the Aurora was "the only organization on which we can count" to combat the efforts of foreign powers seeking "to establish their moral influence over the Chinese."[37] Sarraut acted by reopening the Indochinese University and by reviving the permanent scientific mission in the form of an Institut scientifique d'Indochine.[38]

After Sarraut's departure in 1919, both institutions functioned poorly. The Institut scientifique went out of existence in 1925. The university was held in low esteem, both in Indochina and in France. The director of public instruction for Indochina emphasized in 1924 to Governor General Martial Henry Merlin, himself no fervent believer in higher education, that he had reduced teaching personnel from 142 to 72, saving 41,000 piasters. The director felt that the question of whether to continue the university was no longer on the agenda, as, after all, there were universities nearby in Yunnan and in Hong Kong:

> "Some will accuse me, they have already done it, of megalomania, of dreaming to make a university in the model of the metropolis. You have already sensed, I am sure, the ridiculousness of these allegations. My wish is that a university be made here which responds to the needs of the country."[39]

Exact sciences received little attention. Physics, on the undergraduate level of the P.C.N., was taught to students of medicine, pharmacy, and veterinary science along with science students. In 1928 the professor Bernard, who had been certified in France as *agrégé* in sciences and so assured by the state of a teaching position, counted 95 auditors in an amphitheatre designed for two-thirds that number. He crowed to the director of the medical school that his instruction was the equal of that in France, and that he was certain that many of his pupils "would succeed" in the examination of the P.C.N. *licence* (at the time, perhaps the

37. Aix. Indochine 18375. Consul in Shanghai to Sarraut, 25 Jul 1917.

38. Chassigneux, *L'Indochine* 542-4, for the reopening of the Indochinese University. A. Gruvel in *Actes et comptes-rendus de l'Association Colonies-Sciences* [published as supplementary numbers of the *Revue de botanique appliquée et d'agriculture tropicale*] 13 (1937) 160, for reference to the Institut scientifique d'Indochine, apparently a central directorate for the various observatories and research stations in the colony, one similar to the Institut scientifique chérifien in Morocco. According to the chronology of Edouard de Martonne, reliable in other respects, the Indochinese institute emerged in 1918, three years before the Moroccan one. De Martonne, *Le Savant colonial* (Paris, 1930) 46.

39. Aix. Indochine 51061. Director General of Public Instruction to Governor General, 2 Sep 1924. On Merlin's opposition to advanced education, see W.J. Druiker, *The Rise of Nationalism in Vietnam, 1900-1941* (Ithaca, 1976) 132.

equivalent of an American B.S. in, broadly, physics, chemistry, and natural sciences). Bernard disagreed with critics who claimed "that our advanced instruction is advanced in name only. ... The legend of a 'feeble' university giving 'instruction on the cheap' should go join that of the failure of our school system."[40]

It is difficult to evaluate Bernard's proud comments. Little serious scientific research went on within the university, but, then, the same might have been said about provincial French universities. Professorial salaries were around 10% lower than those of Parisian counterparts — just a bit under what a professor in a provincial university might have earned.[41] With the exception of the school of natural sciences, which had no quotas, places in university programs were strictly controlled according to estimates of positions to be filled by graduates; in 1929 the number to be admitted, exclusive of aspirants in the P.C.N., was fixed at 185.[42] Henri Gourdon, for one, thought in 1928 that it would be good to expand the science school and have it offer serious preparation for the "grandes écoles techniques" of France, and especially begin "the initiation of the natives to scientific research." It would have been far from clear, even to such a strong supporter of higher education as Gourdon, what an *ancien éléve* of a *grande école* could look forward to in his native land.[43]

One wonders whether the relatively poor salaries and unglamorous surroundings could possibly have attracted any talented young physicist

40. Aix. Indochine 51063. Bernard to director of the medical school, cited in Director General of Public Instruction to President of the Chamber of Commerce of Saigon, 26 Sep 1928. Bernard is identified as the professor of physics in Aix, Indochine 51061, course assignments for 1924-25. Bernard is given as *agrégé des sciences* in Aix, Indochine 51136, where he is indicated as sitting on the P.C.N. examining jury for physics and chemistry in 1930.

41. Aix. Indochine 51057. Director General of Public Instruction to Governor General, 5 Feb 1929, where salary scales are compared.

42. Aix. Indochine 51069. Government decree of 5 Aug 1929 specifying the number to be admitted. In "Université Indochinoise: Nombre maximum d'élèves à admettre dans les Ecoles d'Enseignement Supérieure à la rentrée d'octobre 1928", the chart labelled: "Prévisions relatives au nombre des Emplois qui peuvent être attribués par les divers Services aux élèves diplômés de l'Université en août, septembre, octobre, de l'une des années suivantes: 1929, 1930, 1931, 1932" dated 6 Jul 1928.

43. Aix. Indochine 51540. Reform proposals of Henri Gourdon, probably 1928. "Upon returning the engineer trained at the Ecole polytechnique (there have been several) confronts the possibility of a monthly salary much lower, according to a frequently cited example, than that of the French janitor at the University of Hanoi." J. Chesneaux, *Contribution à l'histoire de la nation vietnamienne* (Paris, 1955) 198.

embued with a desire to prepare Vietnamese students for the Ecole polytechnique; in the event, no physicist turned up. The main problem with the Indochinese University, however, remained that its diplomas, not recognized in France, were less well thought of in the colony than foreign diplomas, even those from the Jesuits' Aurora University in Shanghai.[44] Beginning with the first graduating class, *anciens élèves* of Hanoi had to make a special request if they desired that their diploma be granted an attestation of equivalency from Paris.[45] A handful succeeded during the 1920s. One graduated from the Ecole des Mines, three took a *licence-ès-sciences.* Two Vietnamese who finished a *licence-ès-sciences physiques*, Do Tri Le and Hoang Co Nghi, returned to become teachers at Vietnamese institutions. But the number of these highly trained men seems about the same as that in the years before 1914.[46]

Administrative confusion seems to have been endemic to the Indochinese University. It was reorganized in 1925 with schools of law, medicine, fine arts, education, commerce, public works, and agriculture and forestry; seven years later all but the first three schools closed, and coordination of scientific activity fell to the jurisdiction of an unimaginative and powerless Conseil de Recherches scientifiques d'Indochine, which had been called into being in 1928.[47] The imperialist educators seemed fundamentally incapable of creating a separate-but-equal institution of higher learning in the Far East. The suspicion is that they confused education with public instruction.

3b. Higher Learning in Southeast Asia: Dutch Java.

With the beginning of the twentieth century, the colonial regime in the Dutch East Indies put a humanitarian face on its commercial exploitation. An "ethical policy" was inaugurated, according to which the metropolis, in view of the great wealth that it had extracted from the colony, had henceforth the moral obligation to consider above all the welfare of Indonesians. As part of this new turn, Dutch-language

44. Aix. Indochine 51539. Diep van-Ky, "L'Enseignement des indigènes en Indochine", text read to the Académie des Sciences coloniales on 5 Feb 1929. The comparison between the Indochinese University and the Aurora University is on p. 15 of Diep's typescript.

45. Aix. Indochine 51523. Note for the Secretary General, 15 Jun 1921, where the policy is defined.

46. Aix. Indochine 51532 and 51537 for lists of Indochinese students in France.

47. J.S. Furnivall, *Educational Progress in Southeast Asia* (New York, 1943) 87. On the Conseil, de Martonne, *Savant colonial* 46-7.

educational institutions were created or renovated to produce a managerial and administrative elite. Over the first two decades of the century, racial separation ended in higher secondary and professional schools. Education for Indonesians had become a matter of importance, for the relatively small number of colonists lacked the resources to control all facets of their colony's commerce.[48]

The innovations associated with the ethical policy must not suggest that the Dutch had ignored, as the French ignored in Indochina, advanced scientific learning. Colonial Batavia had a scientific society that dated from the end of the eighteenth century. In the nineteenth century were added a museum, a second scientific society, a *gymnasium* for the children of Dutch colonists, and a school for apprising Indonesians of European medicine. Scientific publication was encouraged; the younger society ran a learned journal that equalled any beyond Europe. By 1900, research institutions had taken firm hold. The Royal Magnetical and Meteorological Observatory took its place as one of the world's premier centers of geophysical research. The Botanical Gardens at Buitenzorg were sought out, then as now, for tropical specimens. In 1887 Christiaan Eijkman founded a medical laboratory in Weltevreden, near Batavia; nine years later he and a colleague there established the connection between polished rice and *polyneuritis gallinarum*. This observation led to the discovery of thiamine in the husk of unpolished rice grains and culminated in a Nobel prize for Eijkman.[49]

48. M.C. Ricklefs, *A History of Modern Indonesia* (London, 1981) 143-54. Felix M. Keesing considers the educational systems of French Indochina and the Dutch East Indies to have been of the same form, although it seems to me as if the similarity does not extend easily to the realm of higher learning. Keesing, *Education in Pacific Countries* (Shanghai, 1937) 19.

49. These institutions are treated by a number of articles in the remarkable *Encyclopaedie van Nederlandsch-Indië*, a multivolume work appearing under a succession of editors from 1917. Supplementing the *Encyclopaedie* entries are: L.M.R. Rutten ed., *Science in the Netherlands Indies* (Amsterdam, [1929]); P. Honig and F. Verdoorn eds, *Science and Scientists in the Netherlands Indies* (New York, 1945). The eighteenth- and nineteenth-century background is elaborated in, among other studies: H.A.M. Snelders, "Het Bataviaasch Genootschap van Kunsten en Wetenschappen in de periode 1778 tot 1816", *Documentatieblad Werkgroep 18e eeuw* 41-2 (1979) 62-90; P. Chr. Flu, *The History and Present State of Scientific Medical Research in the Dutch East Indies* (Amsterdam, [1924]); F.D.K. Bosch, "The Royal Batavian Society of Arts and Sciences: A Glance at Its History", *Koloniaal Instituut Bulletin* (Amsterdam) 1 (1938) 116-24; P. Smit, "International Influences on the Development of Natural History in the Netherlands and Its East Indian Colonies between 1750 and 1850", *Janus* 65 (1978) 45-65; Sian Nio Tan, *Zur Geschichte der Pharmazie in Niederländisch-Indien (Indonesien) 1602-1945* (Würzburg, 1976) esp. 62-169; G.M. Lauw, *De Dokter*

Colonial administrators differed about whether the ethical policy meant emphasizing elite education or mass instruction. Christiaan Snouck Hurgronje, the great Dutch scholar of Islam and from 1891 to 1906 a trusted adviser to the colonial government, favored elitism; Alexander W.F. Idenburg, between 1902 and 1919 serving three terms as minister of colonies and one term (1906-16) as governor general, favored basic and practical education for the masses.[50] Yet disagreement over aims did not compromise the steady growth of higher education, a growth entirely stunted by similar debates among French functionaries in Indochina. In 1898 the colonial regime expanded a "Dokter-Jawa" school at Weltevreden, which since 1875 had been running a five-year course of study to train Indonesian medical personnel, by adding a three-year preparatory department in exact sciences and languages; in 1913 a preparatory year was added to the advanced course. The result allowed Indonesian graduates desirous of taking a Dutch medical degree to bypass all theoretical examinations in the metropolis.[51] In 1914 Dutch secondary schools were opened to Europeans, Chinese, and Indonesians without discrimination.[52] University-level education came following Idenburg's replacement as governor general, in 1916, by Johannes Paulus graaf van Limburg Stirum. The new governor's role in the genesis and early evolution of the Technische Hoogeschool (TH) at Bandung, however, was merely one of benign oversight. It is characteristic of Dutch higher learning overseas that the first steps at institution-building were taken by private initiative.

In 1917 a group of traders, industrialists, and agriculturists in the Netherlands conceived of a project to bring higher technical learning to the East Indies. Their project received sanction as the Royal Institute for Higher Technical Education in the Netherlands Indies. They attracted the

Djawaschool: De eerste medische opleiding voor inheemsen in Nederlands Oost-Indië (diss., Univ. Nijmegen, 1979). For the premier astronomer to work in the East Indies during the nineteenth century, L. Pyenson, "Astronomy and Imperialism: J.A.C. Oudemans, the Topography of the East Indies, and the Rise of the Utrecht Observatory, 1850-1900", *Historia Scientiarum* 26 (1984) 39-81; on the foundation of the Royal Magnetical and Meteorological Observatory in Batavia, J. de Bruin and L. Pyenson, "'Gentleman-Scientist': Elie van Rijckevorsel and the Dutch Overseas Effort in Exact Sciences, 1870-1890", *Annals of Science* 43 (1986) 447-73. Of great importance is the volume celebrating the centennial of the Koninklijke Natuurkundige Vereeniging, *Een Eeuw natuurwetenschap in Indonesië 1850-1950* [Jakarta, 1950].

50. Ricklefs, *Modern Indonesia* 148.
51. Flu, *Scientific Medical Research* 10-11.
52. Ricklefs, *Modern Indonesia* 150.

support of the principal Dutch engineering society and of the Ministry of Colonies. By 1919 three-and-a-third million guilders had been collected. The curriculum took shape: Civil engineering would be at its center. The town councilors of Bandung knew how to welcome such an investment: They provided land for a campus. The first building of the TH was in place for inauguration ceremonies in July 1920.[53] The cement guaranteeing firm adhesion of the Dutch creation to Javan soil was provided by a special body, the College of Directors, who mediated between the sponsors in the Netherlands and the curators and staff in the East Indies. The president of the College of Directors, naturally enough, was Java's foremost philanthropist, Karel Albert Rudolf Bosscha.

K.A.R. Bosscha, or "Ru", as he was affectionately known, was born in 1865 in The Hague as the son of the physicist Johannes Bosscha, Jr., professor at the Royal Military Academy in nearby Breda and then director of the polytechnical school in Delft. His mother, Paulina Emilia Kerkhoven, was the daughter of one of Java's oldest tea-plantation owners.[54] Ru's grandfather was a minister of state in the Netherlands. As befitted a professor's son, Bosscha studied engineering at his father's polytechnic. He briefly attended the laboratory of, among others, the physicist Heike Kamerlingh Onnes, and he belonged to a local astronomical association that went by the name of "Copernicus".[55] At the time of his final examination, however, he had a disagreement with one of his examiners. The nature of the controversy is not known; the outcome saw Bosscha leave without taking a diploma. In 1887 he steamed off for the tea plantation of one of his Kerkhoven relatives in the Indies. He floated around for nearly a decade, variously following his

53. The history of the TH is taken from the remarks of R.A. van Sandick, appearing in the official report of the opening of the school, *De Technische Hoogeschool te Bandoeng, stichting van het Koninklijk Instituut voor hooger technisch onderwijs in Nederlandsch-Indië* (Weltevreden, 1920) 3-8, on pp. 3-4.

54. An excellent article has recently appeared on Bosscha, emphasizing his role in supporting astronomy: K.A. van der Hucht and C.L.M. Kerkhoven, "De Bosscha-Sterrenwacht: Van thee tot sterrenkunde", *Zenit* 9 (1982) 292-300. The authors are related to Bosscha, and they have made use of family archives to which I have not been granted access. I have supplemented their description with information from: "Ter Herinnering aan het 25-jarig jubileum van den heer K.A.R. Bosscha als hoofd-administrateur der Assam theeonderneming 'Malabar', Pangalengan bij Bandoeng", *Algemeen Landbouw-Weekblad voor Nederlandsch Indië*, supplementary number (Dec 1920), especially the essay "Een Kwart Eeuw", by C.F. Bodde.

55. "Sterrenkunde (de beoefening der) in Nederlandsch-Indië," in *Encyclopaedie van Nederlandsch-Indië*, vol.VI, 674-9, on p. 676. See also C.H. Hins, "Astronomie" in *Natuurwetenschap in Indonesië* 22-31.

geologist brother Dr. Jan Bosscha on a gold expedition to Borneo, working for a Java telephone company and advising the government, and trying his hand at administering the tea plantation of his relatives. In 1896 his patron Eduard Julius Kerkhoven placed him as director of the Malabar tea plantation. In 1902 Bosscha's geologist brother abandoned Borneo for a neighboring plantation. There they struck it rich. Dividends from Malabar rose steadily from 9% in 1901 to over 80% in 1907. The clever director was quick to diversify his holdings. Bosscha undertook original research to improve his agricultural production.[56] His wealth increased, and so followed his prestige. By the end of the First World War he was one of Java's richest and most powerful scions.

As president of the TH College of Directors, Bosscha — the man who had failed to receive an engineering diploma at Delft — became the power that allowed such diplomas to be awarded by the score. It must be emphasized, nevertheless, that his special attentions turned toward the exact sciences. He was the central benefactor and guiding spirit behind the observatory, also founded privately in 1920, which still bears his name and sits above Bandung at Lembang.[57] We shall see that physics had a special place at the TH, and Bosscha was the one who endowed its laboratory.

There can be no mistake about the aims of the men who created the TH. According to Engineer Rudolf Adriaan van Sandick, a council member of the Royal Institute in the Netherlands, the first step toward university-level education in the East Indies was to be taken by engineering sciences, especially technology. Students at the TH were to become the "general staff" for the army of technicians required to develop the country. To achieve this end, he and his colleagues demanded the highest quality of instruction, inspired by the model of Delft but informed by local circumstances.[58] This principle was revealed in the architecture of the campus — an inspired amalgam of traditional, sweeping West Javan

56. Among the *Mededeelingen van het proefstation voor thee*, published by the Netherlands Indies Departement van Landbouw, Nijverheid en Handel, one finds five papers (nos. 24, 30, 37, 45, and 47) written or coauthored by Bosscha between 1913 and 1916.

57. Van der Hucht and Kerkhoven, "Bosscha-Sterrenwacht". The history of the Bosscha Observatory, especially in its genesis, replicates many of the features of the Bandung TH.

58. Van Sandick, in *Technische Hoogeschool te Bandoeng* 5-6. Van Sandick, during his time in the Indies, had edited the influential journal *Indische Gids*. J. van Doorn, *The Engineers and the Colonial System: Technocratic Tendencies in the Dutch East*

roofs, half-timbered Tudor interiors, and rough stone colonnades.[59] The colonial government also found the TH to be a successful experiment. It watched as enrolment rose from 28 in 1920 to 92 in 1924, the year when the first diplomas and an honorary "doctorate in technical science" were awarded, and when a separate law faculty materialized. It saw the teaching staff rise from 4 to 25, the science professors publishing the results of their new research just as they would have at a university in the Netherlands. When all seemed in order, Governor General Dirk Fock integrated the two faculties into the colonial educational system and assumed responsibility for their operating expenses — some 400,000 guilders per year.[60] The second dean of the Faculty of Technical Sciences was the physicist Jacob Clay, the only professor present at the opening of the TH in 1920.[61] His career, more than that of any other staff member, reveals how the TH incorporated pure learning in its programs.

Clay was born the son of a farmer in Berkhout, near Hoorn, in 1882. His family sent him to the Erasmiaans Gymnasium in Rotterdam, and from there he went, in 1900, to the University of Leiden. Eight years later he received a doctorate under Heike Kamerlingh Onnes for a dissertation on galvanic properties of metals and alloys at low temperatures, but his

Indies [Comparative Asian Studies Programme, Erasmus University Rotterdam, Publication no. 6] (Rotterdam, 1982) 26.

59. The buildings are well preserved in Bandung. Many photographs are in the *Jaarboek der Technische Hoogeschool te Bandoeng, Lustrum-Uitgave 1935* [Bandung, 1935], which also provides a capsule chronology. The Dutch architect, Henri Maclaine Pont, provided his design criteria in his *Concept-Program voor den Bouw der Nederlandsch-Indische Technische Hoogeschool op Java* [The Hague, 1918?]. He began with a collection of barracks, and then transformed it beyond recognition.

60. Funding in: Amsterdam, Koninklijke Akademie van Wetenschappen, *Universiteiten en hoogescholen in Nederland en Nederlandsch-Indië* (Leiden, 1930) 53-102. Here are found comparative statistics on enrolment and funding, and much discussion is given over to educational philosophy. Pious pronouncements from the governor general, the university administrators, the faculty, and K.A.R. Bosscha are found in: Technische Hoogeschool te Bandoeng, *Overdracht der Hoogeschool aan den Lande, op 18 October 1924* [Bandung, 1924].

61. Jacob Clay reported the doctorate of technical science being awarded to Jan Willem Yzerman (a pillar of the Royal Institute for Higher Technical Education in the motherland) in his report as dean of the faculty, *Rede uitgesproken op den vijfden verjaardag der Technische Hoogeschool te Bandoeng* (Bandung, [1925?]) 10. The first earned doctorate, that of Nicolaas Hendrik van Harpen, came in 1930. *Jaarboek, Lustrum-Uitgave* 110. The publications of the TH faculty, taken as a whole, seem to compare favorably, in quantity and in quality, with the learned production of metropolitan colleagues. P.P. Bijlaard, et al. eds, *Publicaties Technische Hoogeschool Bandoeng in de periode 1920-1936* (Bandung, [1936]).

interests extended far beyond experimental physics to embrace Hendrik Antoon Lorentz's theoretical work and the philosophy of neohegelian Gerardus Johannes Petrus Josephus Bolland. Over the next twelve years Clay progressively retired from low-temperature measurements as he forged a philosophy of science based on a critique of neohegelian doctrine. From 1907 to 1920 he taught *gymnasium* in Delft, in 1912 picking up a *privaat-docentschap* in philosophy of science at the institute of technology there. His long essay on the history of the concept of natural law in modern philosophy, in 1915, was crowned with a prize from the Stolpiaans Legaat in Leiden. This and shorter papers in philosophy led to a definitive break with Bolland's brand of neohegelianism in 1919 and, in 1920, to a long treatise on epistemology. When called to become the first professor of physics at the Bandung TH, he was the most prominent philosopher of science in the Netherlands.[62]

The idea of the Dutch curators of the Bandung institute was evidentially to select a physicist who — no rude technologist — would serve to propagate the values of Dutch civilization. Clay's first publications at Bandung were indeed on nature, civilization, and ancient Greek philosophy.[63] As revealing is the fact that the Faculty of Letters at Leiden named him as their first choice, in 1922, to succeed his old mentor Bolland. In their recommendation to the educational authorities in The Hague, however, the university curators reversed the order of preference of the faculty, and a theologically-inclined philosopher from Groningen,

62. J.J. Boasson, "Jacob Clay", *Jaarboek van de Maatschappij der Nederlandse Letterkunde te Leiden* (1955-6) 55-8; G.W. Rathenau, "Levensbericht van Jacob Clay", *Jaarboek der Koninklijke Nederlandse Akademie van Wetenschappen (1955-6) s.v.*; "In Memoriam Jacob Clay", *Synthese* 9 (1955) 423-7; E.W. Beth, "In Memoriam Jacob Clay", *Algemeen Nederlands Tijdschrift voor wijsbegeerte en psychologie* 47 (1955) 23-35; H.F. Jongen, "The Physicist J. Clay", *Synthese* 9 (1955) 428-32; *Biografisch Woordenboek van Nederland*, s.v., also by Jongen. Clay's philosophy of science is elaborated in Klaas van Berkel's study, "Wetenschap en wijsbegeerte in het werk van Jacob Clay (1882-1955)", in: A.F. Heijerman and M.J. van den Hoven eds, *Filosofie in Nederland: De Internationale School voor Wijsbegeerte als ontmoetingsplaats 1916-1986* (Amsterdam, 1986) 65-92.

63. On Clay's neohumanist career at Bandung, see his contributions to the volume *Zeven Artikelen over Grieksche kultuur* (Bandung, 1923) "Het Nieuwe Ontwerp voor het middelbaar onderwijs en de klassieke opleiding", 5-14; "Grieksche Philosophie", 81-9. The latter article was abstracted in the local newspaper, *Preangerbode* (24 Feb 1923). Since Clay introduced the neohumanist volume (a festschrift for the inauguration of a classical *gymnasium*), he is its probable editor. It is remarkable that when the Germans sought a physics professor for the Tung-Chi Universität at Shanghai in 1920, they also found a philosopher of science. Pyenson, *Cultural Imperialism and Exact Sciences* 287-9.

Arthur Joseph de Sopper, assumed Bolland's mantle. Clay and his wife were not unhappy to remain for the time at Bandung. As physicist Adriaan Daniel Fokker (who had inherited Clay's *gymnasium* position and house in Delft and then became professor at the institute of technology there) wrote to Clay a year later:

> "I received the impression from your last letter that you are happy in Bandung (maybe happier than you would have been or be in Leiden?). I believe then that no one has suffered because of the decision of the Leiden curators to change the order of recommendation of the faculty and place De Sopper as # 1?"[64]

Clay was seen as an essential civilizer in the East Indies. Another correspondent wrote in 1922, after the Leiden appointment fell through:

> "With interest I read your speech on nature and civilization and joyfully saw that even if technical education is the real task of the institute, that it also is concerned with philosophy, and that it is moving toward a higher plane than that of merely obtaining practical results. It appeared from your speech that you are the right man for the position you occupy, and you should not regret that the chair in Leiden is taken by another professor."[65]

Clay received the news from Leiden with mixed feelings because he had just inaugurated a new physics laboratory, and his wife Tettje Clasina Clay-Jolles, who was an alumna of physics at Leiden, had gone on the payroll as an assistant there. Tettje explained their quandary in a letter to her sister:

> "It is actually true that we would not like to leave here. Jaap can separate himself from his new lab only with great difficulty, and all in all we have had such a pleasant time here that we would rather stay for another 4 years. Bolland has died an untimely death, as far as we are concerned. But, on the other hand, philosophy is more the love of Jaaps's life, and such an offer will probably never be repeated, and also we do not want to stay forever in the Indies. So we went along with it, but every month that we can stay here longer we consider time won".[66]

The new laboratory, funded by K.A.R. Bosscha, cost some 300,000 guilders, not counting equipment; it came with an instrument maker and glassblower. To inaugurate his laboratory, Clay delivered a discourse on science and civilization (the one referred to in the extract above). He explained that physics was not merely an auxiliary service for

64. Leiden, Boerhaave Museum [henceforth BM]. "Fokker." Fokker to Clay, 4 Mar 1923. Penned postscript to a typed letter.

65. BM. Arch. 234C. Johannes Diderik Bierens de Haan to Clay, 6 Jul 1922.

66. Bilthoven, private archives of Nelke van Osselen-Clay. Tettje Clay to her sister Hesje Jolles, 12 Apr 1922. In 1922 Tettje received 300 guilders per month and Jacob (beginning in July) 1400 guilders per month in salary; "It is absurd that you cannot live [here] on 1200 guilders per month with a family." Tettje Clay to Hesje Jolles, 26 Feb 1922.

engineering. His laboratory would be devoted to research, for physical research provided a fine illustration of how theory and experiment worked together. This was a shining example of the universality of science, in contrast to the case in art and religion, both of which were culture-bound. Pure science, then, was a civilizing force. He intoned, to loud and long applause: "The results of pure thought are the same everywhere. Nothing can further the unity of humanity more than the search for truth."[67]

Which truth was Clay, who had been away from physical research for nearly a decade, to seek? Before steaming off for Bandung, he had had the foresight to spend some time in England with Ernest Rutherford, where he studied about radioactivity and atomic weights.[68] The source of radiation that he came to use was precisely the one attracting researchers everywhere who could not afford a laboratory emitter — the atmosphere. Beginning in 1921 he sent up high-altitude balloons in an attempt to measure a peculiar atmospheric phenomenon observed by many researchers around 1900 and definitely isolated as penetrating radiation (later known as "cosmic rays") by Viktor Hess in 1912. Clay could not get his balloon electrometers to function adequately, but he persisted.[69] Then, around 1925, he took measurements in a deserted hydraulic power station under 83 meters of rock.[70] He presented more high altitude results in 1926.[71] In 1927 the years of patiently perfecting his measuring

67. Technische Hoogeschool te Bandung, *Opening van het Natuurkundig Laboratorium, door Zijne Excellentie den Gouverneur-Generaal van Nederlandsch-Indië, 18 maart 1922* [Bandung, 1922], with addresses by Bosscha and Clay ("Natuurwetenschap en kultuur"), photographs, and diagrams; "De Opening van het natuurkundig laboratorium der Technische Hoogeschool", *Preangerbode*, 18 Mar 1922, which largely reproduces Bosscha's and Clay's addresses. Funding figures from Bosscha's inaugural speech.

68. Jongen, "Clay", *Synthese* 428-9.

69. Clay mentions the 1921 attempts in his article, "Kosmische Ultrastraling", *De Ingenieur in Nederlandsch-Indië* 1, no. 5 (1934) 1-8, on p. 2. He seems to have published his first reports in "Electrische Verschijnselen in de atmosfeer" in: *Handelingen van het derde Nederlandsch-Indisch natuurwetenschappelijk congres* (Buitenzorg, [1924]) 199-206. Atmospheric electricity, it should be noted, had been studied by Dutch physicists in the East Indies for a generation before Clay appeared on the scene. H.R. Woltjer, "De Ontwikkeling der natuurkunde in Indonesië gedurende de 100-jarige werkzaamheid der Koninklijke Natuurkundige Vereeniging" in: *Natuurwetenschap in Indonesië* 32-40.

70. Clay, "Kosmische Ultrastraling" 2.

71. Clay, "Registratie van electrische geleidbaarheid en ionengehalte van de atmosfeer", *Handelingen van het vierde Nederlandsch-Indisch natuurwetenschappelijk congres* ([Weltevreden, 1926]) 128-32.

techniques paid off. Returning to the Netherlands on a year's leave, Clay measured penetrating radiation at various points during the voyage. He concluded that the intensity of the radiation depended on latitude.[72]

The "latitude effect" furnished decisive evidence in support of the view that penetrating radiation, or cosmic rays, consisted of charged particles rather than photons. But at this time the American Nobel-laureate Robert Andrews Millikan had staked his professional career on the assertion that cosmic rays were photons, and that the latitude effect did not exist. Clay moved in 1929 to become professor of physics at the University of Amsterdam. His circle of correspondents and colleagues never doubted his discovery of the effect, but the Americans, having substantial difficulties with their electrometers, hesitated. By 1932 Arthur Holly Compton confirmed Clay's observations. Millikan refused to abandon his views, even after his collaborator Henry Victor Neher reported on the effect in 1933.[73] Clay continued his careful measurements of cosmic rays until the end of his long career. His work was systemically ignored by the Americans and others in their orbit. As Viktor Hess informed Clay with astonishment in 1933, Georges Lemaître and Manuel Sandoval Vallarta attributed the latitude effect to Compton and Werner Heisenberg![74] Compton sought the Nobel prize for Clay, whom he thought shabbily treated by Millikan; he proposed the "initial award to Hess and Clay for the discovery of the rays and the experimental proof that they are extra-terrestrial rays of great energy," and also to Millikan's protégé, Carl David Anderson for his use of the rays to discover the positron.[75] The 1936 physics prize went to Anderson and Hess, in that order. Hess no doubt felt himself lucky, for he had confided to Clay in January of that year: "Millikan is trying his best to make my work appear as insignificant as possible."[76]

72. "Doordringende Straling", *Verslagen Koninklijke Akademie van Wetenschappen Amsterdam* 36 (1927) 1265-77; 37 (1928) 976-83. The reports also appeared in English translation in the academy's *Proceedings*.

73. Ch.C. Gillispie ed., *Dictionary of Scientific Biography*, 16 vols (New York, 1970-80) [henceforth DSB], "Millikan", s.v. D.J. Kevles, "Robert A. Millikan", *Scientific American* 240, no. 1 (1979) 118-25, on pp. 121-2; Kevles's *Physicists: The History of a Scientific Community in Modern America* (New York, 1977), does not refer to Clay although much space is devoted to Millikan and cosmic rays. The Clay-Compton-Millikan controversy is considered in R. H. Kargon, *The Rise of Robert Millikan: Portrait of a Life in American Science* (Ithaca, 1982) 158-61.

74. BM. Arch. 117. Hess to Clay, 16 Jan 1933 and 19 Jun 1933.

75. BM. Compton to Clay, 6 Mar 1937, where Compton describes his good deed of the preceding year.

76. BM. Hess to Clay, 27 Jan 1936.

As Clay flourished in Amsterdam, so his former institute carried on in Bandung. His talented successor there, Herman Robert Woltjer, continued the tradition of physical research. Other research programs came to fruition, such as those of the geodesist J.H.G. Schepers.[77] And it can be said that the prime mover of Indonesian independence, Soekarno, learned about pure science during his years at the Bandung TH from the lectures of one of the most talented physicists of the day.[78]

4. *The South American Missionaries*

Early in the nineteenth century, most of Latin America cast itself free of European colonial masters. But, the same as the United States of America, the new nations continued to live in the shadow of European culture. They were propelled in this direction by the dependence of their economies on the London and Paris financial exchanges and, later, on the policies of enormous US corporations. At the same time, they saw fashions to be apprised of in the Northern Hemisphere, and one fashion concerned the search for special wisdom to aid in transforming the material world. The elites of some new nations — Mexico, Argentina, and Brazil — took natural knowledge in hand and created a self-replicating core of practitioners, eventually absorbing or rejecting foreign-born savants. In other nations, the history of the exact sciences, especially, is a chronicle of foreign interventions and local restorations. The patterns of cultural imperialism noted in Asia are reinforced by considering the presence of French and German researchers — missionaries, I have called them elsewhere — in Ecuador and Chile.

4a. Physicists, Geodesists, and Astronomers in Ecuador

The instauration of exact sciences in middle nineteenth-century Ecuador may be traced directly to the fanatical energy and vision of Gabriel García Moreno. The young García Moreno chose to become certified in law at Quito in the late 1830s and early 1840s, but his real love seems to have been science. He studied under Sebastian Wisse, a resident French engineer, and together with Wisse carried out explorations of the Ecuadorian interior. He visited Europe with his brother in 1850, and then

77. Bijlaard, *Publicaties Technische Hoogeschool Bandoeng* s.v.

78. Ricklefs, *Modern Indonesia* 172-4, 176-80. In *Jaarboek, Lustrum-Uitgave 1935*, Soekarno is listed as having graduated in 1926, but no employment is noted. At the time, he was exiled to the island of Flores after having served a prison sentence for sedition.

returned to Paris in 1855 in order to devote all his energies to the experimental sciences. García Moreno's work attracted favorable comment from his distinguished professors, and upon returning to Ecuador in 1856 he was elected a member of the Geological Society of France. His sympathy for secular learning and French administrative norms was complemented by his deepening conviction as a Catholic. No socialist revolutionary, he came to typify the conservative, nineteenth-century Catholic intellectual for whom science was the revelation of God's truth.[79]

After his second stay in Paris, García Moreno became known as his country's leading scientific intellectual. In 1856 he entered as rector of the University of Quito. The following year he drew up an ambitious plan to reform science instruction, at the center of which was to be an institute of technology. The plan was approved by the senate but lost in the turmoil of politics and wars that engulfed the country. As leader of the Ecuadorian conservative party, García Moreno was elected president in 1861. He ruled for four stormy years during which he conducted an unsuccessful war against New Granada and seriously thought about inviting France to establish a protectorate over his country. He was unable to bring about changes in education. In 1865 García Moreno resigned and went into eclipse. His successors declared war on Spain and, in 1868, saw their country visited by a monstrous earthquake. García Moreno engineered a revolution in 1869 and his election to the presidency in 1870. This time he would have no mistakes. He immediately set about turning Ecuador into a benevolent theocracy.

García Moreno's inspiration, it seems reasonable to assume, was the Jesuit empire in colonial Paraguay. In his view, the Jesuits alone were capable of acting on the decadence and squalor that afflicted his country. He accordingly placed them in charge of reorganizing secondary education and university-level instruction. From his love of pure learning, García Moreno identified science and technology as civilizing forces. The Jesuits would disseminate the claims of natural law while at the same time promoting literacy and public morality. They were to make Ecuador safe for democracy. García Moreno quickly passed by Spain

79. García Moreno, one of the central political figures in nineteenth-century Ecuador, is the subject of numerous studies. His scientific training and interests are admirably recounted by F. Miranda Ribadeneira in *La Primera Escuela politécnica del Ecuador* (Quito, 1972). The present section follows Miranda Ribadeneira's illuminating account.

and France in his search for teachers best apprised of the exact sciences and their applications. As natural knowledge was to build García Moreno's utopia, he handed over the system of higher education in his country to the Jesuit province of Germany.

"The university established in the capital of the State is converted into an Escuela Politécnica," read García Moreno's decree of 1869, "destined exclusively to educate teachers of technology, civil engineers, architects, mechanical engineers, mining engineers, and science teachers."[80] Law and medicine were banished to the provincial *colegios*, or secondary schools.[81] The German Jesuit professors began to arrive in 1870. Juan Bautista Menten, professor of higher mathematics and experimental physics, had been an assistant of astronomer Friedrich Wilhelm August Argelander's at the University of Bonn and a coworker of Pietro Angelo Secchi's at Rome; he was preparing to direct the Jesuit observatory at Bombay when called to become dean and director of García Moreno's new school. Teodoro Wolf, also a former student at the University of Bonn, came as professor of geology and zoology. Luis Sodiro, a native of Italy who had studied at the University of Innsbruck, was the school's botanist. Other mathematicians were José Kolberg and Emilio Muellendorf. And chemistry and geology were remanded to Luis Dressel, also a former student at the University of Bonn who had begun to make a name for himself through his publications. These men were joined in 1872 by mathematicians Armando Wenzel and José Epping and by zoologist Cristian Boetzkes; in 1873 came mathematicians Eduardo Brugier and Alberto Claessen and chemist Luis Heiss.[82] With physics laboratory equipment ordered from the firm of Munich professor Philipp Carl, Quito in short order assembled the finest school of exact sciences in Latin America, one with standards of teaching unsurpassed in the New World. Few if any of the staff had earned doctorates (Wolf received one *honoris*

80. *Ibid.*, 37.

81. In focusing on the dominant role of the German Jesuits in the exact sciences, it is well to record persistent overtures made by García Moreno to French men of learning in law, medicine, literature, fine arts, and elementary instruction. V.M. Rendon, "La France et l'Equateur dans leurs relations politiques et littéraires", Académie des Sciences morales et politiques, Paris, *Séances et travaux de l'Académie: Compte-rendu* (May-Jun 1931) 458-88, on pp. 466-7. Rendon's observations merit caution, as in his assertion that García Moreno asked the French government to establish an astronomical observatory at Quito in 1865.

82. Miranda Ribadeneira, *Primera Escuela politécnica* 48-51, 55, 60. It seems appropriate here to keep the Spanish form of their Christian names.

causa from Bonn in 1877),[83] but well into the twentieth century Jesuit scientists of distinction — the same as Oxford dons — were not required or especially encouraged to go through this particular rite of passage. Their publications, in the event, easily made up for the lack.[84] And they were the least expensive science masters that García Moreno might have found — the total sum paid to the dozen Jesuit professors between 1870 and 1876 being only 60,400 pesos.[85]

As his staff reached critical mass, Menten pushed ahead with plans to erect a major observatory. He ordered a large refracting telescope from Sigmund Merz in Munich; it ran to more than 20,000 francs. He began to construct the observatory building.[86] But all did not go well in García Moreno's Jesuit-run paradise. Jealousies existed between the German Jesuits and those of Spain, the latter of whom had traditionally enjoyed free license in Ecuador. Voices within the Jesuit establishment, including its commandant general, questioned whether it was wise to direct *any* advanced scientific school — Quito being the first experiment of its kind for the order.[87] Most seriously, the German Jesuits at Quito suffered a crisis of faith. Difficulties in setting up his observatory and lack of confidence in his administrative abilities led Menten to leave the order in 1875 while retaining control of the fledgling astronomical institution.[88] He followed Wolf, who had parted ways in 1874 after the ecclesiastical hierarchy in Quito objected to his lectures on Darwinian evolution.[89]

The key to the entire enterprise, the nation's theocrat, might have been able to hold it together, but he fell to an assassin in 1875. With García Moreno's death, the dozen or so Jesuit scientists dispersed. Menten, Wolf, and Sodiro continued on in Ecuador. Wolf eventually published the first great geographical and geological treatise of the country. Menten, who never succeeded in placing the observatory in proper

83. A.N. Martínez, "Sesenta años de recuerdos: El Doctor Teodoro Wolf", Universidad Central del Ecuador, *Anales 287* (Jan-Mar 1934) 181-206, on p. 200.

84. Miranda Ribadeneira, *Primera Escuela politécnica* 365-74, for a list of the professors' publications.

85. *Ibid.*, 301.

86. *Ibid.*, 317, 325.

87. *Ibid.*, 179-81, 191, 204, 206.

88. Menten describes his institution, including its 9-inch Merz refractor, in his *Historia y descripción del observatorio astronómico de Quito* (Quito, 1877). The equipment seems to have been adequate for significant observational work.

89. Miranda Ribadeneira, *Primera Escuela politécnica* 231-2, 245-55 on Menten, 239-41 on Wolf.

working order, went into exile in Colombia during the regime of General Eloy Alfaro; he was hoping to return to Ecuador when death surprised him. The scholarly world (but not the scientific one) has showered its accolades on Father Epping, who in 1881 revealed the subtle harmonies of ancient Babylonian astronomy.[90]

It was on the wreckage of what might have become a German-inspired tradition of learning that General Alfaro opened the second, French chapter of missionary learning in Ecuador. Before he rose to govern the country, he had renovated the Ecuadorian army following the advice of French advisers, and he had sent his son to be trained in the art of French destruction at Saint-Cyr. He then made good use of a French desire to prosecute geophysics.

At the very end of the nineteenth century, French academicians and generals (and in France academicians were not infrequently generals) conceived an urge to measure the meridian arc at a point near the equator, the very task carried out in the eighteenth century by Pierre Bouguer and Charles-Marie de La Condamine. The resulting information allowed one to calculate the extent to which the figure of the globe departed from a perfect sphere — a calculation held to be of use in a number of astronomical, cartographical, and geophysical investigations. The expedition was but one of many overseas and colonial tasks undertaken by the geographical service of the French army. This is not to say that the army was enamored of it. Only after Russia and Sweden mounted a similar expedition to Spitzbergen and after the United States Coast and Geodetic Survey expressed an interest in the Ecuadorian measurement did the French government agree to commit its resources.[91]

The Ministry of Public Instruction planned for the departure of two officers in the geodesical service of the army; they left in 1899 with a budget of 20,000 francs in order to prepare a feasibility study in Ecuador. General Alfaro was lavish in his promises of support. Upon their return,

90. *Ibid.*, 331. On Epping: O. Neugebauer, *The Exact Sciences in Antiquity* (Providence, R.I. 1957) 103-4.

91. G.F. Perrier, "La Mission française de l'Equateur", *Revue économique française* (1907) 441-60, 489-506; Perrier's *La République de l'Equateur: Aperçu général, relations entre l'Equateur et la France* (Paris, 1937) 58-66; and the more extensive description in Perrier's *La Figure de la terre: Les grandes opérations géodésiques* (Paris, 1908) 338-444. The last monograph appeared as volume two of the *Revue de géographie annuelle*. Deep background in: A. Lafuente and J.L. Peset, "La Question de la figure de la terre", *Revue d'histoire des sciences* 37 (1984) 235-54.

the form of the expedition was fixed: The army's geodesical service would carry out the mission, but all scientific questions were to be answered according to instructions from the Academy of Sciences. The expense, some 500,000 francs, was credited to the Ministry of Public Instruction by a parliamentary law of 1900. Additional funding came from philanthropists Raphaël Bischoffsheim and Prince Roland Bonaparte, to which was added Alfaro's commitment for 83,000 francs.[92]

The most striking feature of the expedition relates to its *scientific* side. The Paris Academy seems not to have desired to send a *protégé* of one of its own members as an observer. The distinguished mathematician Henri Poincaré spoke for his academician colleagues when he noted that the geodesical service of the army was perfectly suited for the task at hand:

> "We are not sure of finding their equal; we are sure of not finding any better. We could not expect to otherwise have the cohesion, the familiarity of working together and applying the same methods, and above all the discipline that will permit the work to be carried out quickly and without hesitation."[93]

Any ancillary scientific tasks, Poincaré implied, could be confided to career military men who knew how to follow orders. Poincaré and his colleagues no doubt saw their factotum to be the expedition's geodesist, Lieutenant Georges Perrier, who was the son of the late, first director of the army's geodesical service.[94] It is no reflection on Lt. Perrier, later to rise to the Academy's heights, that the French scientific spirit in Ecuador during the time of the expedition found embodiment in the person of astronomer François Gonnesiat.

Gonnesiat, forty-four years old in 1900, had logged an unimpressive record as a researcher and teacher. A pupil of Academician Charles André's, he found himself marooned at Lyons as an *aide astronome* and *chargé de cours*. When Ecuador's General Alfaro sought to profit from France's military presence by calling a French astronomer to reorganize the observatory at Quito, the position fell to Gonnesiat. His tenure at Quito lasted until 1906, that is, for precisely as long as the French army was traipsing around the Ecuadorian countryside.[95]

92. AN. F-17 13062. General J.A.L. Bassot to Minister of Public Instruction, 4 Oct 1900 and 29 Aug 1900; Bonaparte to Bassot, 10 Jan 1900; Bischoffsheim to Bassot, 29 Jan 1900; Joseph-Hippolyte Frandin to Delcassé, 27 Feb 1901; Perrier, *Figure de la terre* 355.

93. Poincaré, cited in Perrier, *Figure de la terre* 346.

94. "Perrier, Georges", *DSB* s.v.

95. Information on Gonnesiat's career comes from the various editions of the

When Gonnesiat arrived in Quito, he found something akin to a museum. The observatory still had Menten's nine-inch Merz equatorial and a six-inch Repsold meridian telescope, and around 1890 it had benefitted from an input of more than 48,000 marks, but, as Gonnesiat noted, "most of the instruments deteriorated before having been put to use."[96] Just as Gonnesiat began his tasks, General Alfaro ceded to General Léonidas Plaza Gutiérrez in a general election. Plaza was less of a francophile and patron of science than his predecessor had been, and the new man promptly cut back on Gonnesiat's resources. The French astronomer protested to influential senators and intervened with Plaza, but he was pessimistic about the long-term outcome: "It is to fear that he [Plaza] considers geodesy and astronomy as a bit superfluous in a country that lacks a great many things." As far as the operation of his observatory went, Gonnesiat felt saddled with a "quite mediocre and hardly conscientious" support staff. To place himself in the best light with the local community, he emphasized, he lectured three times a week at the university, and he invited his fifteen pupils to carry out practical work in the observatory.[97] The university left much to be desired, and a good deal of the blame for this situation had to be taken by the secondary schools: "The level of secondary instruction in science is so low that it is necessary to begin everything anew at the faculty." In general, students bypassed science in favor of law and medicine.[98] Gonnesiat noted grimly that the faculty of science had even ceased to operate for five years.[99]

In a letter of 1901, Gonnesiat signaled that the French military expedition was also having difficulty adjusting to General Plaza's regime.[100] The organizational problems are clearly revealed in a report tabled by the expedition's leader in 1902. The leader, Commandant Robert Bourgeois, like Perrier later to be received into the Academy of Sciences, emphasized that Alfaro had failed to keep his promise of providing the mission with 50,000 francs for the year 1901. In a confidential part of his report, dated February 1902, Bourgeois identified the source of the problem. It was the French *chargé d'affaires* in Quito, Joseph-Hippolyte

standard biobibliography, *J.C. Poggendorff's biographisch-literarisches Handwörterbuch*. When no reference is provided for biographical information in the present text, it may be assumed that "Poggendorff" is the source.

96. AN. F-17 13062. Gonnesiat to Director (of the Paris observatory?), 23 Dec 1900.

97. *Ibid.*, Gonnesiat to General Bassot, 19 Nov 1901.

98. *Ibid.*, Gonnesiat to Director (of the Paris observatory?), 23 Dec 1900.

99. *Ibid.*, Gonnesiat to General Bassot, 14 Jan 1902.

100. *Ibid.*, Gonnesiat to General Bassot, 19 Nov 1901.

Frandin. He had systematically obstructed the work of the mission. Bourgeois identified the lack of cooperation in Frandin's wounded pride: Frandin felt slighted at not having been asked to be the official head of the military mission, and he resented not having been named an officer of the Legion of Honor for his efforts with regard to the mission. In addition to undermining the geodesists, Bourgeois emphasized, Frandin had been neglecting to watch out for a German military mission stalking the halls of the Ecuadorian Ministry of War.[101]

In 1906 the French military missionaries returned to Paris with their data, which Perrier worked up in desultory fashion over the next thirty years. Gonnesiat, his unimpressive term as observatory director near an end, was able to persuade the local authorities to establish a French-staffed school of sciences. To this end, in 1905, four French teachers signed a five-year contract with the government at Quito. The new arrivals had barely unpacked their bags when a political revolution occurred. Three of the four men promptly voided their contract and returned home, leaving only the chemist, Gentey, to begin teaching.[102]

The change in government might or might not have argued for Gonnesiat's return to France. His Parisian masters settled the matter. The Ministry of Public Instruction assigned him to succeed Charles Trépied as director of the Algiers observatory. Having promoted the civilizing mission of France in South America, he graduated to a similar capacity in the colony of Algeria.[103] His post at Quito fell to Lyons astronomer Joanny Philippe Lagrula. After two years, Lagrula left to become *astronome adjoint* at the new Nice observatory.[104] Here the standard histories of French science in Ecuador end. There was, nevertheless, an important sequel that reveals how Parisian mandarins sought to use exact sciences as a cover for achieving larger, geopolitical goals. The sequel concerns the secret mission to Ecuador of a newly naturalized French citizen, Milan Ratislav Štefánik.

The son of a Protestant pastor at the village of Kosariská in Austrian-

101. *Ibid.*, Robert Bourgeois, "Compte-rendu sommaire des travaux exécutés par la mission de l'Equateur en 1901" (Paris, 1902).

102. Perrier, *Figure de la terre* 385; J.M. Vargas, *Historia de la cultura ecuatoriana* (Quito, 1965) 394-5. Vargas presents an inaccurate picture of the French *savants*.

103. AN. F-17 13582. "Inspection annuelle" (Algiers), 10 Mar 1908.

104. R. Baillaud, *Souvenirs d'un astronome (1908-1977)*, vol. III (Rodez, France, 1980) 35.

controlled Slovakia, Štefánik studied exact sciences at the institute of technology and at the Charles University in Prague, as well as in Zurich. He took a doctorate from the Charles University in 1904 and then went to Paris with the hope of becoming an assistant to Jules Janssen at Janssen's astrophysical observatory in Meudon. Introduced to Janssen by Camille Flammarion, Štefánik worked at Meudon until he was forced out by Janssen's jealous lieutenant and successor, Henri Deslandres. Štefánik, while proud of his Slovakian roots, preferred to make a name for himself in the French cultural sphere. He became a roving astrophysicist trouble-shooter for the French astronomical community, undertaking a great many missions in France and abroad during the years before 1914. Having become a French citizen in 1912, he served as a French officer in the war, rising from lieutenant to, in 1918, general. Among his military exploits rank his service in the French air corps and with the French army on the Italian front. In 1918 the French promoted him to general and charged him to investigate the status of Kolčak's Czech legion in Siberia. Having been a member of the Czech-Slovak Council in Paris, he became the first Minister of War in the independent state of Czechoslovakia. Štefánik died when the plane returning him from Italy to Bratislava was fired on by Slovak troops.[105]

Štefánik became involved with Ecuador through a scheme conceived by the undersecretary of the merchant marine, Anatole de Monzie. In 1913 Štefánik sought to return to Tahiti, where in 1910-11 he had set up an astronomical station in conjunction with his observations of Halley's comet and a solar eclipse. Senator Emile Chautemps, in whose house Štefánik was a regular visitor, thought to send the astrophysicist to the South Seas through the new Panama Canal. De Monzie was charged with the arrangements. The civil servant had been thinking hard about the imminent opening of the American-controlled canal, and how to secure a French-controlled coaling station on the Pacific side. Clipperton Island, a French dependency claimed by Mexico, was too far north of the main route to French Oceania. For the preceding decade, however, the Galapagos had been much on the mind of French civil servants.[106] De

105. An enormous literature was generated on Štefánik during the interwar period. Short accounts may be found in L. Mrkosová-Pajdusáková, *M.R. Štefánik, astronom* (Bratislava, 1970); Josef Ihnat. *et al.* eds, *General Milan R. Štefánik 1880-1980: Historical Profile* (New York, 1981).

106. W. Clement, "Die Monroedoktrin und die deutsch-amerikanischen Beziehungen im Zeitalter des Imperialismus", *Jahrbuch für Amerikastudien* 1 (1956) 153-67. E.T. Parks and J.F. Rippy, "The Galapagos Islands: A Neglected Phase of American

Monzie recalled that one of the Galapagos had become the property of a Basque expatriate, who upon his death willed the island to the small town of Barcus in the Basses-Pyrénées. Štefánik thought to present his travel plans to de Monzie along the lines of erecting a radio transmitter on Tahiti. De Monzie identified the French-claimed Galapagos island as an essential relay station for Štefánik's project.[107]

De Monzie and Štefánik hatched their plans in the course of the next week. Štefánik would be granted a mission by the Ministry of Public Instruction to the observatory at Quito. He would gain the confidence of the Ecuadorians by his scientific expertise, following which he would obtain for France a wireless concession. De Monzie successfully negotiated for credit from his ministry, from the Geographical Society of Paris, and from the steamship lines. Štefánik left for Ecuador in September 1913.[108]

Arriving in Quito, Štefánik found the observatory almost in ruins. The roof leaked, and "in every corner there were spiders." By the end of November, Štefánik had helped the observatory's director, Luis G. Tufiño, place the Merz equatorial and the meridian circle in working order. Then he announced to his new friends his plan of erecting an *astronomical* station on the Galapagos. He wrote in his Slovak diary at this time:

> "Situation: England will want to exercise its rights as soon as a foreign power occupies the Galapagos. Its interest is that the Galapagos remain under the jurisdiction of Ecuador. On the other hand, the United States wants the Galapagos to support Panama, and also, I think, because of the imperial utopia inaugurated by Roosevelt. If the United States does not succeed in purchasing them, it will succeed by other means. Ecuador will declare war for the Eastern Territories and

Strategy Diplomacy", *Pacific Historical Review 9* (1940) 37-45. Around 1900, President Eloy Alfaro of Ecuador claimed that France offered him 100 million francs for a free port in the Galapagos.

107. A. de Monzie, *Destins hors série* (Paris, 1927) 127-53.

108. See Štefánik's otherwise incomprehensible grant proposal submitted to the Geographical Society and his equally bizarre proposals to the Ministry of Public Instruction for setting up a permanent meteorological and astronomical station on Tahiti (refused) and for meteorological and wireless research in the Galapagos (accepted). Paris, Bibliothèque Nationale, Archives de la Société de Géographie de Paris, colis 76, no. 4306; Štefánik to Director of the Society, 17 Aug 1913 and 20 Aug 1913. AN. F-17 17286, Štefánik's two proposals to the Ministry of Public Instruction. The Ministry of Colonies awarded Štefánik 7000 francs toward the cost of his Tahitian expedition. SOM. Océanie C 16, carton 104. Minister of Colonies to Governor General of French Oceania, 31 Jul 1913.

will occupy the islands and sell them the United States, or else the United States will send agents to the islands who will foment a revolution and acknowledge a republic, as in Panama."

Štefánik continued, several days later, with a succinct statement of his strategy:

"It is in the interest of France that Ecuador be francophile, which would put an end to the expansionism of the United States toward the south. Ecuador needs money and culture. France would be inclined to a discussion, because it wants new commercial possibilities since investing in Germany would be suicide. But financial experts do not want to enter adventurously into this republic. And Russia, without consolidation of its debts, does not give enough guarantee. The Chinese government is pressuring French politicians about a certain financial arrangement. I came to explore the terrain. Scientifically, my trip is sufficiently motivated to be above all suspicion. And, besides, our intentions are in reality friendly. It is a pity that my health and the scientific hopes do not let me carry through my intentions and my tasks. I think that he [Reinbourg] is the better mediator for carrying through this arrangement between France and Ecuador. And thus, I could return as soon as possible. But I will not return empty-handed. I must have the wireless; and I will also leave a great scientific program for Quito."

Reference is made to his comrade in arms, a Dr. Reinbourg, who was attached to the French embassy, and to a delicate ulcer that had already undergone an operation in Paris. The immature, romantic, and frenetic adventurer who is revealed in these lines, however, was perfectly suited for the task at hand. Štefánik persuaded the Ecuadorians to sign with the French instead of with Telefunken. He engineered a new plan of activity for Quito's observatory which required purchasing French equipment and sending Tufiño to Paris for advanced training. The last note in his diary indicated: "All fixed up ... I did not like the diplomatic chore, and I feel gastric pains from time to time."[109]

The secret mission, serving to advance "French expansion," propelled Štefánik into the Legion of Honor, but not much else concrete resulted from it.[110] No powerful transmitter came to Tahiti, hence no relay station to the Galapagos and no French maritime intervention in Ecuador. Štefánik corresponded with Tufiño, urging him "always to remain loyal to your second country that you call France," for it was "the duty of France to make all necessary efforts to transform your beautiful country into a powerful, rich, independent, and cultured state;" there is no indication, however, that the Ecuadorian managed to travel to Paris or

109. All quotations from Mrkosovà-Pajdusáková, *Štefánik* 142-146. I am grateful to George Franko for help in translating the Slovak text.

110. Quotation in the citation for the Legion of Honor. *Ibid.,* 147.

that French equipment was sent to Quito. At the beginning of the war, German ships were openly received at Ecuadorian ports, and the Germans ran a wireless relay station in the Galapagos. De Monzie asserted, nevertheless, that during the war Štefánik's mission of cultural imperialism was "more useful than our tracts and warlike orders" in keeping Ecuador friendly to the French Republic.[111]

4b. Astronomers and Geophysicists in Chile

In Chile, as in Ecuador, French and German men of learning dominated the exact sciences during the four generations following independence from Spain. The intervention of foreign savants in the southern nation, however, was less episodic than in the case of Ecuador. Competition between French and German researchers may not have resulted in especially original publications, but there can be little doubt that, by around 1920, the exact sciences had become fixtures of Chilean intellectual life.

French talent dominated the scene in the first decades after independence. At this time, Chilean authorities were above all concerned with producing an inventory of their country, and a topographical map ranked first on their list of priorities. French geographers, following in the wake of revolutionary armies, had set the standard for cartography, and so Chilean eyes turned to Paris. Charles Ambrose Lozier, a thirty-eight-year-old French geographer living in Buenos Aires, was engaged in 1823 as a teacher for the city of Concepción (Lozier had been trained in the pedagogical methods of Joseph Lancaster) and charged to produce a map of Chile. After four years he effectively abandoned his geographical work; he worked as an engineer and ended up as a hermit. The Chileans tried again to recruit a French *savant*. In 1828 they procured the services of Claude Gay as teacher of physics and natural history in the French-staffed *colegio* founded at Santiago by Pierre Chapuis; in 1830 he received a contract to study the natural history of Chile. Gay spent the next nine years collecting information in Chile, toured Peru, and in 1842 returned to Paris, where he died in 1873. His labors resulted in a general survey of Chile, published in 1855.[112]

111. *Ibid.*, 149. De Monzie, *Destins hors série* 146, citing a letter from Štefánik to Tufiño of 13 Sep 1914; global evaluation on p. 151.

112. E. Greve, "Don Amado Pissis y sus trabajos geográficos y geológicos en Chile", *Revista chilena de historia y geografía* no. 107 (Jan-Jun 1946) 71-170, on pp. 72-9. Greve's sensitive study provides an excellent introduction to the history of exact sciences in nineteenth-century Chile.

The last of the heroic French geographers, Aimé Pissis, arrived in 1848. Born in the Haute Loire in 1812, he studied at the Ecole polytechnique and the Ecole des Mines. In 1837 he took part in a French geological expedition to Brazil; he returned to France five years later. A second South American voyage began in 1847, to Bolivia's interior, where he measured the heights of various mountains. His work being cut short by political events, he was planning to return to France in 1848 when he received a call from Chile to direct a geological and mineralogical survey of the country. Pissis devoted the next thirty years of his life to the project. He produced detailed maps and geological studies, in the process garnering membership in the French Legion of Honor. In his case the honor seems superfluous, for he was no special agent of French political designs. He attracted the best European assistants he could find, both French and German.[113]

As if to balance Gallic geology, Chilean authorities took advantage of the revolutions of 1848 to call in a German astronomer. External events propelled them to do so. In 1849 the United States navy sent an expedition to Chile to observe the retrograde motion of the planets Mars and Venus and, in conjunction with Northern Hemisphere measurements, thereby obtain a precise measure of the solar parallax. The naval officer in charge, James Melville Gilliss, had previously been delegated the task of planning the Naval Observatory in Washington. Gilliss based his Chilean research program on the calculations of Marburg astronomer and polymath Christian Ludwig Gerling. The Americans passed three years at Santiago de Chile, making measurements. Their activity attracted the attention of Chilean president Manuel Montt and Montt's educational adviser Ignacio Domeyko. The Chilean government offered to purchase the expedition's equipment for 8,000 pesos and use it to establish a permanent observatory.[114] Beginning in 1850, three Chilean

113. *Ibid.*, 82-3, 107-8. Among Pissis's assistants were Friedrich Wilhelm Döll, a surveyor educated at the Cassel polytechnic (p. 137), and Martin Drouilly Laurent, a French engineer (pp. 142-3).

114. Astronomy in Chile is treated in: F.W. Ristenpart, "Astrónomos alemanes en Chile", 177-93 of *Los Alemanes en Chile*, the first volume of the festschrift of the German Scientific Society of Santiago, published for Chile's independence centenary in 1910; R. Grandón M., "El Observatorio astronómico nacional: Renseña de su desarrollo y actividades en sus primeros cien años de existencia, *Anuario del Observatorio astronómico nacional de la Universidad de Chile* (1952) 7-35; Ph. C. Keenan, S. Pinto, H. Alvarez, *The Chilean National Astronomical Observatory (1852-1965)* (Santiago de Chile, 1985). Pp. 17-52 of the last volume, which contains identical texts in English and in Spanish, concern the period before 1923. The Gilliss expedition

assistants worked with Gilliss, and one of the three, Gabriel Izquierdo, was detached from Pissis's geographical survey team.

It is significant in the present context to consider why the Chileans did not think to have the observatory directed by one of Pissis's countrymen. At this time, President Montt and Domeyko turned to their principal scientific advisers, the brothers Philippi. Bernhard Eunom Philippi was a German naturalist-turned-entrepreneur who, through the middle of the nineteenth century, vocally supported German colonization in southern Chile. Rudolf Amandus Philippi directed the polytechnical school at Cassel; when political changes forced him out in 1850, he went to Chile, the promised land of his younger brother.[115] Seeing the advantage in signing on a student of Gerling's to continue Gilliss's work, Bernhard Philippi enthusiastically recommended one to his friend Salvador Sanfuentes Torres, the administrator of the southern province of Valdiva who had since gone on to become secretary general of the university in Santiago.[116] In 1850 twenty-five-year-old Karl Wilhelm Moesta arrived as Gilliss's heir apparent. Until Gilliss left, Moesta received an appointment, ironically, as an assistant to Pissis.

Moesta assumed direction of the observatory with Gabriel Izquierdo as an assistant, although Izquierdo soon left for the Chilean Corps of Engineers.[117] Moesta felt the need for a German-trained coworker. He asked the Chilean consul general in Hamburg, who was then acting as an agent for colonization, to find one. An offer went out to another of Gerling's students, who declined in favor of a German post. In 1855 the

is considered in all three treatments. For the internal, political dimension: W.D. Rassmussen, "The United States Astronomical Expedition to Chile, 1849-1852", *Hispanic American Historical Review* 34 (1954) 103-13; J. Harrison, "Science and Politics: Origins and Objectives of Mid-Nineteenth Century Government Expeditions to Latin America", *ibid.* 35 (1955) 175-202.

115. G.F.W. Young, *Germans in Chile: Immigration and Colonization, 1849-1914* (New York, 1974). German colonization is exhaustively treated in J.P. Blancpain's *chef d'oeuvre: Les Allemands au Chili (1816-1945)* (Cologne, 1974) [*Lateinamerikanische Forschungen; Beihefte zum Jahrbuch für Geschichte von Staat, Wirtschaft und Gesellschaft Lateinamerikas*, vol. VI], although Moesta makes no appearance in it. A summary account, where science makes no proper appearance, is available in G.J. Ojeda-Ebert, *Deutsche Einwanderung und Herausbildung der chilenischen Nation (1846-1920)* (Munich, 1984) [*Beiträge zur Soziologie und Sozialkunde Lateinamerikas*, vol. XXX].

116. Greve, "Pissis" 94.

117. Here I follow *ibid.*, 92-5 ff., 139, 143-4, supplemented by the *DSB*, s.v., on Hansen and Schumacher and "Poggendorff", s.v., on Volckmann.

consul succeeded in hiring Hermann Volckmann, at a salary of a bit more than half that given to Moesta. Volckmann was such a catch that Pissis offered him a raise and captured him for the geographical survey. Moesta looked again to Germany. In 1859 he received, on the recommendation of the Gotha observatory director Peter Andreas Hansen, Richard Schumacher — the son of the director of the Altona observatory Heinrich Christian Schumacher (who had been Hansen's teacher and patron). Once more Pissis wooed the new astronomer to the staff of the geographical survey. For a time around 1859 Moesta employed Volckmann's brother Friedrich in an honorary capacity; Friedrich Volckmann went on to become associate professor at the University of Marburg.

Despite the perennial problem of finding trained assistants, Moesta pursued a vigorous and innovative program in observational astronomy at the finest research observatory in the Southern Hemisphere. He relocated the observatory to a more suitable site and oversaw construction of new buildings. He mounted expeditions to observe solar eclipses. He observed the declination of Mars at its 1862 opposition, in cooperation with his predecessor Gilliss, who had risen to direct the Naval Observatory in Washington; the result provided a new value for the solar parallax, and hence the distance of the earth from the sun.[118] By the middle 1860s, Moesta grew weary. His teaching responsibilities at the university in calculus and astronomy (with twenty-odd students) sapped at his energy. In 1865, he applied for leave to attend an astronomy congress in Leipzig. He never returned. At issue may have been the terms of his contract, as well as the destruction by shipwreck of a new 24-cm refractor that he had just ordered. The terms of his departure were amicable, for beginning in 1869 he acted as Chilean consul general in Dresden. In 1875 his catalogue of over 2000 southern stars appeared with a Dresden press.[119]

Moesta's successor was José Ignacio Vergara, an assistant whom he had

118. Keenan, *et al.*, *Chilean National Observatory*, 19-21 discusses the international project, which involved nine observatories, in detail.

119. C.G. Moesta, *Observaciones astronómicas hechas en el Observatorio Nacional, 1856-1860*, (Dresden, 1875). This publication appears cited in Chile, Ministerio de Educación Pública, *Exposición retrospectiva de la ensenanza: Obras de los profesores, catálogo* (Santiago de Chile, 1941) 143; C. André and A. Angot, *L'Astronomie pratique et les observatoires en Europe et en Amérique, depuis le milieu du XVIIe siècle jusqu'à nos jours*, part 4: *Observatoires de l'Amérique du Sud* (Paris, 1881) 15; Grandón, "Observatorio astronómico", 13. Compare Keenan, et al., 24.

trained. Vergara's ambitions, however, lay in the domain of politics, and he rose to occupy important ministerial portfolios. He presided over the steady deterioration of Moesta's old operation. Finally, in 1883, he asked Moesta to find two astronomers for Santiago. Moesta complied. Adolf Marcuse arrived as the "first astronomer", followed shortly thereafter by Wilhelm Wickman as the "second astronomer." Marcuse had obtained a doctorate at Berlin under Adolf Hirsch; he was a veteran of the 1882 German expedition to observe the transits of Venus, and he had spent some time observing at Pulkovo in Russia. Wickmann seems never to have published a scientific paper. The two Germans were at each other's throats from the beginning.[120] Marcuse provided information about the observatory's poor condition which appeared in a book criticizing Vergara's management of the institution; Vergara promptly fired him. He returned to Germany where, in 1907, he finally received an associate professorship at the University of Berlin. Wickmann resigned soon thereafter to direct the Quito observatory from 1887 to 1895.[121]

The unhappy experience with the two young German astronomers persuaded Vergara to look to France for replacements. Jean Albert Obrecht, Irénée Lagarde, and Javier Devaux arrived in 1888 as first, second, and third assistants.[122] Vergara died in 1889, and Obrecht assumed his post as director. Obrecht was a native of Strasbourg who, after his city went over to the Germans in 1870, continued to be educated in French system. He obtained a doctorate from Paris in 1884 for observing the satellites of Jupiter, and had several years' professional experience at the Paris observatory under his belt when he steamed off for Santiago. Lagarde and Devaux returned to France in 1890, as Chile careened toward civil war.[123] Obrecht stayed and survived the retribution exacted in 1891 from the war's losing side. For the next fifteen years, Obrecht was unable to mount a research program. He failed to carry out his contribution to the international astrographic catalogue — a project conceived and directed by French astronomers — and he published no other observations. He watched while the Americans from Lick Observatory in California mounted two major expeditions to Chile, one to the eclipse of 1893, the other a long term commitment beginning in 1903 to observe radial velocities of southern stars. The radial-velocity

120. Moesta's advice on the candidates and the candidates' incompatibility according to Ristenpart, "Astrónomos alemanes".

121. This from Keenan et al., following Ristenpart, "Astrónomos alemanes".

122. *Ibid.*, 29.

123. Grandón, "Observatorio", 116.

astronomers, belonging to the "D. O. Mills Expedition," set up their 92-cm mirror on a hill overlooking Obrecht's observatory; they overshadowed the Chilean institution in every way.

Obrecht's tenure as director had seen the continual rise of Germany's prestige in Chilean affairs. During the civil war of the early 1890s, the dominant role was exercised by a German officer who had been reorganizing the Chilean general staff. The officer, Emil Körner, recast the army in a German mold and over the next fifteen years sent dozens of Chileans to study war in his native land.[124] In the 1880s, Chilean educator Valentín Letelier argued for a national normal school, along the lines of German pedagogical seminaries. He finally persuaded the minister of education to hire six German professors as the core of his Pedagogical Institute. Their contracts were initially set for six years, and their annual salaries ranged from 1000 to 2800 pesos. Along with the senior professors came dozens of positions in the secondary schools. Among the professors, over the period before the First World War, were a number of mathematicians and physicists, such as Richard Poenisch and August Tafelmacher. Resources for research were denied them. In their pedagogical institutions, physics was considered to be a descriptive science, where mathematics had no part; yet no provision was made in the early 1890s and through at least 1903, at the Pedagogical Institute, for a physics laboratory.[125] By 1913 Germans were by far the most numerous among the foreigners teaching in Chilean public institutions; in the teaching corps, the number of native German speakers approximately equalled the number of native English and French speakers combined.[126]

Obrecht acted to counter the strong German threat by helping found, in 1891, the Société scientifique du Chili. It was to be a learned academy

124. F.M. Nunn, *Yesterday's Soldiers: European Military Professionalism in South America, 1890-1940* (Lincoln, Nebraska, 1983) 102-6. Blancpain, *Allemands au Chili* 702-50.

125. Blancpain, *Allemands au Chili*, considers the invasion of the German pedagogues in detail, pp. 666-701. On pp. 772-3 he provides a list of the first thirty pedagogues hired in 1889-90 for the Pedagogical Institute and for the secondary schools. Contracts are considered on p. 675, the situation of physics laboratories on p. 693.

126. The entire teaching corps of Chile, their nationalities, assignments, and salaries, are provided in: Chile, Ministerio de Instrucción pública, *Personal de instrucción pública: instrucción superior, secundaria i especial* (Santiago de Chile, 1913). Listed are 49 Germans, 23 French, 21 English, 16 Italians, 12 Spaniards, and 10 Swiss, among other nationalities.

supported by the patronage of the French colony in Chile. The initial officers were drawn from the small number of French professors in Santiago (with Obrecht presiding), and the society's council drew on the talent of commercial and professional émigrés. The founders — and only French citizens could qualify for the title of "Founding Member" — hoped that the society would act to bring French ideas to Chile by holding monthly meetings, establishing a library, and publishing learned proceedings and memoirs. The task in Chile was great, Alphonse Nogués emphasized in his presidential address in 1892. Only a single real laboratory, that in chemistry run by one of the Germans, existed in the country; the physics collection at the university was "a beautiful museum containing numerous and beautiful apparatuses carefully kept in the glass cabinets, tightly closed off and spotlessly clean." French academicians sent encouragement. Pierre Puiseux, an astronomer at the Paris observatory, wrote to Obrecht in 1892 about the honor that would accrue from having Chileans "consider France as their natural instructor." Such "moral influence" was especially important, in Puiseux's view, given that France had little enough leverage there.[127]

Obrecht and his colleagues devoted many years of their lives to the society, but their ambitious plans eventually failed. Despite the fact that the proceedings appeared bilingually beginning in 1892, by 1896 membership had stalled at around 200, and attendance at the society's sessions never passed ten. Ambitious, not to mention pretentious, monographs appeared in the society's *Notes et mémoires*, but the level of discourse descended to the ridiculous. Obrecht published a text on celestial mechanics, and the associate director of the Santiago natural-history museum Fernand Lataste launched an extended polemic arguing against Darwinian evolution, but these exercises in high science rubbed elbows with a long-running debate on solutions to the problem of the trisection of an angle.[128] The society soon lost two of its prominent supporters, as Lataste returned to France in 1892 after the death of his

127. Santiago, Société scientifique du Chili, *Actes* 1 (1891) xx for the society's French charter and character, 2 (1892) xxvi for Nogués remarks, 2 (1892) xcviii for extracts from Puiseux's letter. Nogués, an *Ingénieur des mines,* was professor of industrial physics and technology at the University of Santiago. M. Vega E., *Album de la colonie française au Chili* (Santiago, 1904) 90.

128. J.A. Obrecht, "Sur une nouvelle méthode de détermination des orbites des planètes et des comètes", Société scientifique du Chile, *Notes et mémoires* 3 (1893) 65-104, 171-205. In a number of articles from this period, Lataste challenged the "transformist" observations of the French naturalist Rémy St. Loup. The trisection debate began in the *Actes* 3 (1893) lxxxviii, xcvi-xcviii, cxviii-cxix.

wife and two of his children and when Noguès died in 1895.[129] By 1896 funds no longer existed to continue the ambitious publications schedule and to set in order the growing collection of books and periodicals that had been arriving as gifts and as a result of corporate exchanges. The society went into eclipse behind German-inspired culture.

Given Obrecht's mediocre performance as an astronomer and as a guiding light of the Société scientifique, it is not surprising that he was eased out of his position when, in 1906, Pedro Montt became president of Chile. Like his father Manuel Montt, the new president was partial to German culture. He accelerated the German direction of higher learning by calling a German astronomer to replace Obrecht (who nevertheless continued teaching at the university) in directing the Santiago observatory.

The new man, Friedrich Wilhelm Ristenpart, arrived in 1908. He had obtained a doctorate from Strasbourg in 1891 and for the next nine years moved from one assistantship to another, finally landing a position as scientist with the Prussian Academy of Sciences in 1900. He advanced to become privatdocent at the University of Berlin in 1904. When he left Berlin, he had had the foresight to obtain a two-year leave of absence from the Prussian Kultusministerium, his employer, although shortly after arriving in Santiago he felt quite optimistic about spending the rest of his life there. Ristenpart brought in two younger men to serve on his staff, Walther Zurhellen and Richard Prager. Zurhellen had a doctorate from Bonn and Prager one from Berlin; like Ristenpart, both Zurhellen and Prager were experienced and published astronomers. The three men set to work, Zurhellen on the celestial-chart and Prager supervising the calculations. Ristenpart had a broader vision. He wanted to relocate the observatory to a better site, obtain more modern instruments, and secure better salaries. He found the site and supervised the transfer of the observatory; he ordered the equipment, especially a 60-cm refractor from Grubb, Parsons & Co.; he continued to publish; but he failed to provide for his expanded staff.[130]

President Pedro Montt died in 1910, soon after having arrived in Germany for medical treatment. Montt's successor was not merely

129. Vega, *Album* 82-3, 90.

130. In addition to the general histories of astronomy in Chile, see Prager's obituary of Ristenpart in *Vierteljahrsschrift der astronomische Gesellschaft* 49 (1914) 14-22, which seems to be the *Urquelle*.

ambivalent about astronomy: In 1912 he froze appropriations for the new observatory site, and initiated an audit into alleged financial improprieties. Ristenpart found opposition from his large staff. His style was perceived as autocratic and inflexible. After he publically humiliated Zurhellen, the latter resigned in 1912; Zurhellen found a position at the Berlin-Babelsberg Observatory, and he died while fighting in France in 1916. After offering Zurhellen's position to a number of candidates, a student of Johannes Hartmann's at Göttingen, Franz Pingsdorff, accepted. The furor surrounding Ristenpart's administration, having reached the press, led the government to cancel his contract and have him replaced by the lawyer and editor who had negotiated his hiring, Enrique Matta Vial. Ristenpart, having in only a few years restored confidence in Chilean astronomy, was mortified. Several days after Vial took charge of the observatory, Ristenpart committed suicide. He left a son by his first marriage and a wife and three daughters by his second. He did not live to see the birth of his third daughter.[131]

It took the government less than two months to call back Professor Obrecht as scientific director of the observatory. The curatorial administration was entrusted to an assistant-director confidant of the government's, Ismael Gajardo Reyes. It was a power-sharing arrangement that Ristenpart had explicitly rejected in his last months, but Obrecht was content to return to the observatory in a diminished capacity. The notion of *cumul*, where a scientist or scholar held positions at a number of institutions, was a French tradition that had sunk deep roots in Chile. The position in astronomy carried a large salary, and with it Obrecht became one of the highest-paid academics in Chile.[132] That he was selected to succeed Ristenpart is testimony to the strong support that he had cultivated during his years of lecturing at the School of Engineering and Higher Mathematics of the university.

Obrecht had come to a point in his life when astronomical research meant little. Certainly he was uninterested in continuing commitments under-

131. Keenan, et al., *Chilean National Observatory* 39-47. The authors provide a fine discussion of Ristenpart's administration; their treatment must be seen as more accurate than my own preliminary survey in "Missionary Diffusion of Exact Sciences: Functionaries and Seekers in Latin America, 1850-1930", *Quipu: Revista latinoamericana de historia de las ciencias y de la tecnología* 2 (1985) 372-415, on pp. 412-3.

132. *Personal de instrucción pública*, 167, records his salary as being paid in sterling, not (as with most teachers and professors) in pesos or (as with several *savants* of foreign origin) in gold. The figure amounted to 25,150 pesos. The rector of the university, Solar Domingo Amunátegui, received a salary of 27,855.35 pesos (p. 92).

taken by his predecessor. He argued in favor of abandoning the new site for the observatory which had been chosen by Ristenpart, already partially constructed. He abandoned Ristenpart's plans for preparing charts of the southern skies down to the tenth photographic magnitude, and he was unenthusiastic about continuing Santiago's role in the international astrographic catalogue project. His staff eroded. Pingsdorff, recruited by Ristenpart and having arrived in Santiago the day before his director's suicide, found life difficult under Obrecht — especially after the beginning of the war. Obrecht forced him to move out of his house and gave it over to Gajardo, the administrative chief; Pingsdorff resigned in 1915. Richard Wüst, Ristenpart's mechanic, alone among the Germans remained under Obrecht's regime. Obrecht suffered a stroke in 1922. Gajardo, who had risen in 1916 to become associate "scientific" director, succeeded him. There were no funds for hiring new staff, and until he was relieved in 1929 Gajardo ran the observatory without much concern for its scientific program.

The German influence remained strong, nevertheless, in the person of Walter Alfred Knoche. In 1910 Ristenpart succeeded in severing the meteorological service from its long-time dependency on the observatory. Knoche was called in 1910 to direct the service, which in its new form was called the Instituto central meteorolójico y jeofísico. He had recently completed a doctorate at Berlin under meteorologist Wilhelm von Bezold, and had served as assistant at the Prussian meteorological institute and the weather office in Berlin. Like Ristenpart, he was an accomplished and dedicated researcher. Knoche trained local assistants and published a portion of his results in Spanish. In one of his major undertakings, he worked with Jakob Laub, Albert Einstein's collaborator who was then professor of geophysics at the National University of La Plata in Argentina, on an extensive magnetical and meteorological study of the Brazilian eclipse of 1912.[133] After a decade in Chile, Knoche's interests turned toward climatology and the connection between environment and civilization. In 1919 he became director of an institute for medicine, technology, and public health in Santiago. He moved to Madrid in 1932, only to return to Chile in 1936. He finally went to work for the Argentine national meteorological service, and in 1945 he died in Buenos Aires.[134]

133. Pyenson, *Cultural Imperialism and Exact Sciences* 201.

134. "Walter Knoche", *Geofisica pura ed applicada* 8 (1946) 1901. "Poggendorff", vol. V, appendix, s.v.

The French approach to exact sciences in Chile resulted in geographical chronicles, rather than more abstract calculations or observations. Obrecht was uncomfortable with the kind of exact sciences that had emerged late in the nineteenth-century, which demanded that observation, experiment, and theory be interrelated; Ristenpart flourished under such a regime. Obrecht was content to run the observatory as, if not an institution for sinecured *érudits*, at least a government office responsible first to the practical concerns of the nation; Ristenpart took charge of a factory for producing new knowledge, one where the employees were worked ruthlessly. The counterpoint of French and German styles of scientific activity is seen here with striking clarity. The work of a final French scientist in Chile, Fernand, comte de Montessus de Ballore, provides a subtle variation on the theme.

Fernand Jean Baptiste Marie Bernard, comte de Montessus de Ballore, received his training at the Ecole polytechnique. In 1879, at age twenty-eight, Artillery Captain Montessus was charged with a military mission to San Salvador. There his interest in seismology was awakened. He returned to become a teacher at his alma mater, and in 1888 he published his first major statement on seismology, an analysis of earthquakes and volcanoes in Central America. Feeling that he lacked sufficient information for making synthetic statements in the field, he began a systematic inventory of historical earthquakes. The inventory, left in manuscript form when he died in 1923, was to be his life's work, coming to occupy around 30 meters of shelf space. The great Valparaiso earthquake of 1906 led the Chilean government to establish a seismological service, and the authorities called Montessus de Ballore to direct it. At the time, German and British geophysicists were taking enormous strides in the science of seismology, the analytical and observational foundations of which were laid in the years before 1914; the Chileans saddled themselves with someone who had strengths in neither mathematical analysis nor instrumentation. Montessus, a latter-day natural-history fact collector, lectured at the University of Santiago on construction in earthquake-prone areas. He continued to work at extracting broad generalizations from his catalogue until the very end of his life. His last book, a treatise in French on seismology, was edited by his brother Robert. No disciple followed the path that he took.[135]

135. Ch. Davison, *The Founders of Seismology* (Cambridge, 1927) 160-76; A. Renier, "Fernand-Jean-Baptiste-Marie Bernard comte de Montessus de Ballore" in: Montessus de Ballore, *La Géologie sismoloqique* (Paris, 1924) 453-8.

It is natural to find such contrasting styles in Santiago at this time, because nineteenth- and early twentieth-century Chile was, more than any other Latin American country, a stage for competing German and French interests. A substantial French colony existed there — 16,000 to 18,000 people — to oppose the waves of German immigrants.[136] All could see the contrast, although interpretations about its meaning varied. Following a visit to Bolivia in 1895, for example, André Bellesort commented on differences between French and German immigrants. The French, in his view, represented "unintelligent individualism" and desired immediate personal gain; the Germans, however, sought to consolidate their colony and think of the superior interests of their "race." Bellesort emphasized: "One group [the Germans] say let us conquer a new province; the other group, let us amass enough to return quickly to our own land."[137] Given these competing styles, and the metropolitan energies spent to mobilize them, it must be considered a triumph of Chilean statesmanship for the country to have remained neutral throughout the First World War.

5. *Incursions into the British Empire*

The world colored pink in 1914 — to this day the British Empire is shaded in pink on historical maps — was connected to its metropolis by a network of coaling stations, transcontinental railroads, and transoceanic telegraph lines. Conquered peoples were to deal with imperial councillors in London through the distilled representations of their chief, local regent, or (in the case of Quebec) provincial first minister. Order was kept by the threat of imperial troops conveyed in naval squadrons from one quarter of the globe to another. Culture was to be instilled by scores of institutions of higher learning, small variations on metropolitan forms which were staffed by metropolitan émigrés or the descendants of British settlers. It is doubtful if most of the world's pink peoples spoke much English or thought of petitioning the king. They probably worried a bit about the thin red line that might obliterate their community. And, from a familiarization with the various colonial schools that sprouted at

136. H. Lorin, "Impressions du Chili: Les Chiliens et la France", *Revue des deux mondes* 7(1912) 192-215, on p. 210; J.P. Blancpain, "Francisation et francomanie en Amérique latine: Le Cas du Chili au XIXe siècle", *Revue historique* no. 544 (Oct-Dec 1982) 365-407.

137. A. Bellessort, *Le Jeune Amérique: Chili et Bolivie* (Paris, 1923) 325.

administrative and commercial centers, local intellectuals would have retained respect for metropolitan factories of knowledge.[138]

Over the past decade it had become commonplace to emphasize the slow evolution of self-sustaining scientific communities in the countries of the British Commonwealth.[139] Much space has been given over to identifying national characters and styles. It is not my intent here to join that discussion. Rather, I consider how rival European scientific communities tried to intervene in directing the course of science at two locations in the British Empire where they believed support would materialize from the conquered peoples — French Canada and Afrikaans South Africa.

5a. France Patronizes French Canada

In 1920 the province of Quebec contained some two million French-speaking inhabitants, making it the largest French community beyond Europe. In Quebec, a tradition of serious scientific research was represented at English-language McGill University in Montreal, where William Osler and Ernest Rutherford had begun their stellar careers.[140] No comparable tradition existed at French-language Laval University in Quebec City and at French-language medical and engineering schools in Montreal.[141] In the French sector, education at all levels remained under the control of several orders of the Catholic Church. After fifty years of rapid expansion, Quebec's polyglot metropolis, Montreal, was also the commercial capital of Canada. It was the juncture where rail lines from the west met ocean-going ships. English-speaking financiers controlled the fortunes of the city and the destiny of the Dominion. The Quai d'Orsay in Paris sent scientists into this inhospitable territory as agents of French culture. Their entry was facilitated by a special, bilaterally-funded institution which had quarters in the newly founded University of Montreal.[142]

138. The anglocentrism of the Empire is revealed in the scientific journals preferred by physicists in the colonies: L. Pyenson and M. Singh, "Physics on the Periphery: A World Survey, 1920-1929", *Scientometrics* 6 (1984) 279-306.

139. There are veritable industries in place now for documenting the twists and turns of scientific enterprise in Canada and Australia, to mention only two cases with which I am familiar.

140. A survey is provided in S.B. Frost, *McGill University: For the Advancement of Learning*, 2 vols. (Montreal, 1980-84).

141. A profound, comparative assessment is provided in Y.M. Rabkin and J.A. Lévi-Lloyd, "Technology and Two Cultures: One Hundred Years of Engineering Education in Montreal", *Minerva* 22 (1984) 67-95.

142. The finest study of the nineteenth-century background to science in French

The University of Montreal emerged in 1918 when a branch of Laval University acquired a charter from the religious authorities in Rome. The new institution was presided over by a faction of the bourgeois and ecclesiastical elite of French Canada who sought above all to populate the liberal professions. By all accounts, secular scholarship and scientific research remained foreign to the desires of the founders. Far more important was to assemble a faculty who were devoutly Roman Catholic and practiced in the art of lecturing. The pool of local talent being insufficient, university administrators looked abroad for staff. They did not think to fathom the French university system for candidates, nor to inquire with the state universities of Belgium — much less that at Geneva. Only European Catholic institutions would do. For the post of biology professor, French Catholics identified Louis-Janvier Dalbis to the *Canadiens*.

At the time that he was recruited by the university rector, Dalbis sat as an undistinguished thirty-nine-year-old teacher at the Collège Stanislas in Paris. The latter, elite secondary school was closely affiliated with the Marianist Fathers and claimed, as a staff member, the philosopher Jacques Maritain.[143] Dalbis was not in Maritain's league; although he had authored several teaching manuals, he seems never to have published an original scientific paper. When called to Montreal he — the same as nearly all his new colleagues — lacked advanced certification; in his case, he did not have the *doctorat-ès sciences*. It took him nearly five years to settle into his tasks in the new land. He received as salary 3500 dollars and a travel bonus. Having the service of a demonstrator paid at around 1500 dollars to relieve him of unpleasant chores, and without the least interest in himself advancing the frontiers of knowledge, Dalbis had time on his hands. He was active in setting up the Montreal Société de Biologie and in founding a similar forum for other high-minded citizens — the Association canadienne-française pour l'Avancement des Sciences (ACFAS). Dalbis realized his principal contribution to scholarly life, however, in his role at the Institut scientifique franco-canadien (ISFC).[144]

Canada is C. Pighetti's *Scienza e colonialismo nel Canada ottocentesco* (Florence, 1984) esp. pp. 165-202. The role of provincial politics, early in the twentieth century, is considered in R. Duchesne, *La Science et le pouvoir au Québec* (Quebec City, 1978). The early years of the University of Montreal are sketched in O. Maurault, *L'Université de Montréal* (Montreal, 1952). The history of recent French influence in French Canada is provided in L. Roussel, *Les Relations culturelles du Québec avec la France 1920-1965* (diss., Laval Univ., 1983).

143. Collège Stanislas, *Le Collège Stanislas (1804-1979)* (Paris, [1979]) 14.

144. Roussel, *Relations culturelles*, 47-50, 82-4, 138-40, and *passim*. R. Rumilly, *Le Frère Marie-Victorin et son temps* (Montreal, 1949) 89, 132.

As the ISFC was Dalbis's brainchild, he became its first president. The aim of the institute was to bring in illustrious professors from France for short lecture tours in Montreal and the rest of Canada, to send French-Canadian professors to the Sorbonne (the implication being that they would teach French-Canadian studies there), and to organize special scientific "missions" of a nature mutually agreeable to French Canada and France. Lawyer Edouard Montpetit, secretary-general of the university and vice-president of the ISFC, emphasized that the new institute was a clearing house, for it had neither buildings nor professors of its own. From its foundation in 1926 until 1931 it found lodgings in the rooms of the University of Montreal; thereafter it found refuge in the Ecole polytechnique, the school that served in lieu of the university's lacking engineering faculty.[145]

In the eyes of ISFC vice-president Montpetit, the institute had been called into being to contribute directly to the technological development of French Canada. He wanted French scientists to bring "their methods, their ways of doing things" to Montreal. "We need French science, French methods," he exclaimed.[146] The whole operation was "to serve the expansion of French culture."[147] The problem with this formulation, however, was a lack of symmetry. French Canada had no illustrious scientists to send to France. Then, too, there was the matter of the propagation of the faith, a consideration that weighed heavily against inviting, say, Jewish or Marxist professors from France. It is typical of the situation that the medievalist and philosopher Etienne Gilson — a travelling propagandist for French culture overseas — came as the first French envoy, although one senses from remarks made in the opening session of the ISFC that this choice was perceived as slighting the desideratum for visiting professors to lecture on the natural sciences.[148]

Dalbis had originally envisioned McGill's Department of Romance Languages as a party to his institute. McGill refused to contribute to this cause, but its administrative head, General Sir Arthur Currie, turned up at the ISFC's inauguration.[149] Dalbis did manage to send his distinguished

145. Montreal, Archives de l'Université de Montréal, Fonds Institut scientifique franco-canadien [henceforth UM]. P12/A1. E. Montpetit, "L'Institut scientifique franco-canadien", n.d.

146. *Ibid.*

147. UM. P12/A1. "Rapport sur l'Institut scientifique franco-canadien", n.d.; P12/E3. French Consul General in Montreal to Dalbis, 20 Mar 1928.

148. UM. P12/A1. Document without title.

149. UM. P12/E1. McGill University to Dalbis, Apr 1926; Dalbis to McGill University, 6 May 1926; Principal of McGill to Dalbis, 18 Jan 1927.

visitors to the city's premier university. As he wrote to McGill physicist Albert N. Shaw: "The speakers of the Franco-Canadian Scientific Institute will always be very happy to go to McGill University and to speak there as often as possible."[150] When the institute brought in Léon Brillouin for a short visit, McGill physicists were glad to have him talk. And another French physicist missionary, Georges Dejardin from Lyons, was elected an honorary life member of the McGill Physical Society.[151] There was, of course, much showmanship involved in these exchanges between English and French institutions. Arthur Stewart Eve, professor of physics at McGill, was quick to invite Dalbis to Werner Heisenberg's lectures when Heisenberg passed through McGill in 1929.[152]

Regarding the other French-language university in North America, Laval University, Dalbis was more circumspect. Its scientific pedigree was longer than the University of Montreal's, and it resented the expenditure of provincial monies for the ISFC at Montreal. Laval scientists seem to have concentrated their lobbying on the Association canadienne-française pour l'Avancement des Sciences, an avowedly nationalistic pressure group that rejected French intrusion into Quebec's affairs. The ACFAS found a vocal supporter in Dalbis's principal rival at the University of Montreal, the botany professor and director of the botanical gardens, Brother Marie-Victorin. The two biologists fell out over the ISFC and over provincial politics.[153] Dalbis was, at any rate, obliged to inquire at Laval on behalf of his missionaries. Some made the trip north; others, like Brillouin and Jean Cabannes, did not. As Abbé P.J. Filion, rector of Laval, demurred to Edouard Montpetit regarding Cabannes: "We don't believe that we can find an audience sufficiently well prepared for this sort of talk."[154]

The ISFC acted quickly, in the late 1920s, to attract scientists of note. In 1926 experimental physicist Aimé Cotton declined Dalbis's invitation, although he left open the possibility of a future engagement.[155] Dalbis tried unsuccessfully to get mathematician Robert Fernand Bernard, vicomte de Montessus de Ballore (the brother of the Chilean seismologist

150. UM. P12/E1. Dalbis to Shaw, 19 Mar 1929.
151. UM. P12/E9. Arthur Stewart Eve to Dejardin, 18 Mar 1932.
152. UM. P12/E4. Eve to Dalbis, 20 Mar 1929.
153. M. Lavallée, *Marie-Victorin: Un Itinéraire exceptionnel* (Saint-Lambert, Quebec, 1983) 225. Marie-Victorin called the ISFC "a real betrayal" of ACFAS.
154. UM. P12/E5. Filion to Montpetit, 14 Sep 1929.
155. UM. P12/E1. Cotton to Dalbis, 12 Dec 1926.

Fernand) who taught at the Catholic University of Lille.[156] Dalbis also invited Henri Villat from the Institute for Fluid Mechanics in Paris. Villat, an expert applied mathematician, was studying aerodynamics for the French air corps and could not come.[157] He would in any event have received a mixed reception. As Dalbis complained to the archbishop of Montreal later, Villat's French lectures seemed "a bit too advanced" for the students of Quebec.[158]

Dalbis's efforts in exact sciences were first gratified by accident. In 1927, Léon Brillouin, the brilliant French theoretical physicist, went on a lecture tour of the United States, and he wrote to Dalbis that he wanted to spend ten or so days visiting French and English universities in Canada. He was willing to speak on the new quantum physics, atomic structure, statistics, or telecommunications.[159] Dalbis leaped at the opportunity. Brillouin came to Montreal and lectured at its two universities; he was flattered by Dalbis's attentions.[160] More successful was Dalbis's catch in Jean Cabannes, a physicist from Montpellier. According to Cabannes's plans, during his ten-week stay in North America, he was to give one lecture on the radiation of energy, meet three times with those doing research work in laboratories, and give twelve lectures on the physics of radiation at the Ecole polytechnique. His talks at McGill, and in the rest of Canada, would be added on to this charge.[161] Dalbis wrote on Cabannes's behalf all across Canada, especially dunning the French consul in Vancouver. Although Cabannes was a young man, Dalbis wrote to the consul, he would "bring to Canadian universities the latest discoveries of French physics." For this reason, Dalbis asked for the consul's financial and moral support in sending Cabannes to the Pacific.[162] Dalbis was delighted with Cabannes's triumphant tour, congratulating him on "the accomplishment of a mission as useful to Canada as to French scientific expansion."[163]

Cabannes's successful tour led, in 1932, to that of physicist Georges Dejardin, a thirty-nine-year-old *professeur titulaire* of physics at the University of Lyons. Dejardin, who was travelling in any case to Harvard

156. UM. P12/E6. Montessus de Ballore to Dalbis, two letters of 9 Jan 1930.
157. UM. P12/E6. Villat to Dalbis, 10 Sep 1928 and 22 Dec 1928.
158. UM. P12/E7. Dalbis to Paul-Napoléon Bruchési, 25 Jul 1930.
159. UM. P12/E2. Brillouin to Dalbis, 23 Dec 1927.
160. UM. P12/E3. Brillouin to Dalbis, 1 Apr 1928.
161. UM. P12/E5. Institut scientifique franco-canadien to unknown, 10 Aug 1929.
162. UM. P12/E4. Dalbis to French Consul in Vancouver, 27 Sep 1929.
163. UM. P12/E6. Dalbis to Cabannes, 14 Jan 1930.

and the Massachusetts Institute of Technology, also spoke on the new physics and radiant energy.[164] He planned to provide some forty lectures of one hour each:

> "15 talks on radiation, the new physical theories, the optical properties of X rays, visual and physical photometry; 10 talks on photoelectricity, with an accent on photoelectric cells and their applications (phototelegraphy, television); 5 talks on thermionic phenomena (wireless valves, rectification of alternating current, etc.); 3 or 4 talks on research and several popular expositions."

In the last presentations, he would go into radiation theory, industrial applications, and high-altitude atmospheric research. Dejardin really wanted to lecture on the level of advanced instruction in France, but he asked Dalbis if that were appropriate.[165] As Dalbis did in Cabannes's case, he wrote to Canadian universities to host Dejardin; some claimed poverty, others declined to engage a physicist who could not speak English.[166] Dalbis was nevertheless successful in having the French consulate in Vancouver contribute 6000 francs for Dejardin's voyages west.[167] The French physicist worked hard for his money. He gave 27 talks in the United States and 54 in Canada.[168]

All this cost a great deal of money. During the 1920s the government of the province of Quebec contributed 6000 dollars annually, and the French Ministry of Foreign Affairs gave the same amount in francs. A measure of the value that France placed on its Montreal enterprise is obtained from its continual renewal of the ISFC grant, even in the depths of the depression when it reduced funding of all other "French works abroad" by ten percent.[169] The money from Quebec was placed in deposit, in dollars, at the Canadian National Bank in Paris; the French contribution went to the same bank in francs.[170] For the daily functioning of the institute, the University of Montreal contributed about $1000 in terms of matériel and personnel, and occasional private donations ran to about $500 annually. Half of the Quebec monies went for honoraria to the invited professors; the rest financed travel inside Canada and the

164. UM. P12/E8. Dejardin to Dalbis, 6 Nov 1931.
165. UM. P12/E8. Dejardin to Dalbis, 21 Oct 1931.
166. UM. P12/E9. Toronto claimed poverty: Robert Falconer to Dalbis, 9 Feb 1932; the University of Western Ontario raised linguistic considerations: Sherwood Fox to Dalbis, 9 Feb 1932.
167. UM. P12/E9. French Consul in Vancouver to Dalbis, 10 Mar 1933.
168. UM. P12/E2. ISFC council minutes, n.d.
169. UM. P12/E10. Dalbis to Athanase David (Secretary of the Province), 21 Nov 1933.
170. UM. P12/A1. E. Montpetit, "L'Institut scientifique franco-canadien", n.d.

salaries of the institute's administrators. Cabannes, for instance, received $1000 from Quebec, Dejardin only $500, and Brillouin $100. The ISFC's agent in Paris, Etienne Gilson, dispensed francs for transatlantic travel and, one thinks, some of the salary of the invitees. Gilson also received a handsome award for his services.[171]

When Dalbis — under pressure from Quebec nationalists — returned to France in 1932, he administered the ISFC's accounts. He was well paid, in 1936/37 receiving $2000 for his troubles.[172] With his repatriation, however, the train of physicist missionaries to Quebec stopped. The distinguished visitors left no real mark on science in French Canada.

5b. Dutch Astronomers in South Africa

At the beginning of the twentieth century, Afrikaans South Africa, like French Canada, was populated by European settlers many generations removed from their origin. It may be argued that, in both cases, the settlers pushed out indigenous peoples to set up an essentially self-sufficient agrarian regime that was politically conservative, piously fundamentalist, and uncritically anti-intellectual. By the end of the nineteenth century, the regimes had been eroded by British capital investments in the natural-resource sector of the economies — especially mining but also timber and hydroelectric power in French Canada and foodstuffs in South Africa. The Boers fought an unsuccessful and bloody war to remain independent; the Dutch saw them as potential if reluctant allies, much as the French saw the *Canadiens français*.

Dutch astronomers penetrated South Africa in a number of steps, and at each step their government weighed out its encouragement, to the extent that it did, in grams. The pattern is similar to the one present in the East Indies. The Dutch government required evidence of strong metropolitan interest and local financial support before it would consider funding permanent institutions of higher learning.

The earliest long-term contact with astronomers in South Africa emerged at University of Groningen. There, in 1878, twenty-seven-year-old Jacobus Cornelius Kapteyn arrived as professor of astronomy. The state, however, would not provide its astronomer with a telescope. Kapteyn contacted David Gill, the director of the Royal Observatory in Cape

171. UM. P12/1F2. "Etats financiers", 1926/7 to 1940.
172. *Ibid.*

Town, offering to measure the plates that Gill had taken of the southern sky. The resulting long-distance collaboration, between 1886 and 1896, brought Kapteyn's "astronomical laboratory" to world attention. He visited South Africa only after this project had essentially been completed.[173]

The second step taken by Dutch astronomers in South Africa, while also a private venture, proceeded in a completely different manner, for its prime mover had a special (not to say idiosyncratic) temperament. The protagonist, Joan George Erardus Gijsbert Voûte, was in fact a central figure in Dutch overseas astronomy during the first part of the twentieth century.

Voûte was born an eldest child in Madioen, Java, in 1879, his parents having married the year previously in Batavia. Joan's father Cristoffel descended from a long line of devout Hugenot exiles. Following the custom of Dutch colonists, Voûte's parents sent him and his brothers to the Netherlands for their education. They lived with a Welsh grandmother and attended *gymnasium* in Amsterdam. Voûte then went to the Delft Institute of Technology, and in 1908, at the rather late age of twenty-nine, graduated a civil engineer. At a time when practitioners of exact sciences flashed through university like blazing meteors (Kapteyn was one such meteor) Voûte came to his life's work at a leisurely and measured gait. His mother, Maria Antoinnetta de Dieu Stierling, had died in 1896, and his father followed in 1904. Since arriving in the Netherlands, he had become a virtual orphan, although he did see his mother just before she died. An inheritance freed him from the vagaries of the work-a-day world. While studying at Delft he discovered astronomy and the objects of his heart's desire. These were double stars. Upon leaving Delft, he went to the Leiden observatory to train as an astronomer.[174]

Voûte worked at Leiden for two years without pay before being named, in 1910, the "third observer" there. This lowest professional position in the observatory hierarchy (coming after the director, the first observer, and the second observer) had been created especially for Antonie

173. *DSB*, "Kapteyn", s.v.

174. D.J.K. O'Connell, "Joan George Erardus Gijsbertus Voûte", *Quarterly Journal of the Royal Astronomical Society* 5 (1964) 296-7. Voûte family archives, Amsterdam, and information from the Walburgh Schmidt family, formerly of The Hague.

Pannekoek in 1896. Voûte saw before him several paths to follow. He could make a career of observing and hope to rise to the position of first observer, as his superior Jan Hendrik Wilterdink had just done. Or he could leave the ladder at any point for other parts. In 1878 Kapteyn had resigned as second observer to become professor of astronomy at Groningen; in 1906, Pannekoek had resigned as third observer to work in Berlin as an astronomer and socialist educator.[175] Voûte stayed at Leiden for five years. An ambitious observer would have begun to publish and would certainly have completed a doctorate in astronomy. This was not, however, Voûte's way. He married Anna Lorch and continued his craft out of love. By 1913 it should have become evident to Voûte that he would never be able to rise to a commanding position in astronomy in the Netherlands. His gaze turned outward. A mark could be made in the Southern Hemisphere. With money from his mother-in-law, he purchased equipment and set off for South Africa.[176]

Voûte's destination was the Royal Observatory at the Cape of Good Hope, then under the direction of Sydney Samuel Hough. Voûte remained at the Cape for five years, carrying out his double-star survey of the southern skies. If his stars remained constant in their course, his domestic life oscillated from one extreme to another. He divorced Anna Lorch in 1915, the next year marrying Cornelia Wilhelmina Ulrich van Noorten. He divorced for a second time in 1917. With no regular position, and without support from wealthy in-laws, Voûte fell back on his own, private resources. These would probably have been sufficient insofar as mere existence was concerned, but Voûte had come to desire a permanent place in the world of astronomers. In 1918 he wrote to Willem de Sitter, professor of astronomy at Leiden, asking to be rehired. The expatriate astronomer had heard that the director of the observatory, Ernst Frederik van de Sande Bakhuyzen, had died and that the first observer had quit; he asked De Sitter, as the senior astronomer, to think of him when the observatory was reorganized. Voûte no doubt knew that De Sitter had spent several years at the Cape under Hough's predecessor, David Gill.[177] When De Sitter took over the observatory from Van de Sande Bakhuyzen in 1919, however, there would be no place for Voûte.

175. W. de Sitter, *Short History of the Observatory of the University at Leiden, 1633-1933* (Haarlem, 1933) 44-5.

176. Information from the Voûte family archive in Amsterdam and the Walburgh Schmidt family formerly of The Hague.

177. Leiden, archives of the Leidse Sterrewacht [henceforth LS]. Voûte to De Sitter, 2 Jul 1918. Voûte family archives, Amsterdam.

The latter may or may not have had a realistic understanding of his chances at Leiden. He did turn most of his efforts toward being appointed lecturer in South Africa.

Voûte's campaign centered around finding a post in astronomy at the Afrikaans-language University of Stellenbosch, a local college elevated to university status in 1918. The university council decided to offer instruction in astronomy, but it balked at purchasing expensive equipment. Voûte asked De Sitter to help the cause "in word and deed" by lending the university a 7-inch refracting telescope and one of the old universal instruments that Leiden had. Once these were in place, he believed that donations would be forthcoming for bigger instruments. "It is difficult to convince the authorities" about the need for new equipment, Voûte emphasized. "Most of the council members are farmers and don't give a hoot about it ... But support from the professor at Leiden university and observatory, and lending instruments, will count for a great deal."[178] De Sitter saw no drawbacks to Voûte's plans: "The Leiden observatory shall with great pleasure support these efforts and sees it as its duty to do what it can during the first difficult years." He would look into lending the two desired instruments. If things turned out well at Stellenbosch, more equipment would be forthcoming.[179]

Voûte waited into 1919 for his lectureship, but money for it did not materialize. He then purchased some land near Stellenbosch and resolved to erect a small observatory of his own, using his 5-inch Zeiss camera. He asked De Sitter whether he could not himself be recipient of Leiden's astronomical largesse; he would be happy with a small salary and the promised equipment.[180] De Sitter gave serious consideration to Voûte's new plan. The Leiden curators, he wrote, had approved lending instruments to an institution in a friendly country, but it was not certain if they would agree to placing the instruments privately. Neither was the Leiden theoretician sure about entering Voûte on the payroll. Yet it might be possible, he thought, to appoint Voûte as an assistant in charge of the instruments. This would amount to at best 2000 guilders per year, with a small subsidy for purchase of photographical plates and erection of the telescopes. Since De Sitter had to present a budget to the curators in the near future and could not hope to obtain more precise information

178. LS Voûte to De Sitter, 10 Aug 1918 and 26 Aug 1918.
179. LS De Sitter to Voûte, 9 Jun 1918.
180. LS Voûte to De Sitter, 12 Jan 1919.

from Voûte before then, he decided to "take a gamble" and include provision for the arrangement in his proposals.[181]

Voûte continued seeking a position at Stellenbosch, but he also looked to Johannesburg and the Union Observatory there. His letters from the middle of 1919 appear frenzied, the writings of a middle-aged man going through a personal crisis. Voûte did not want De Sitter to think that he was only interested in securing a position: "Yes, you know that I am doing it out of love for our science."[182] Stellenbosch remained unresponsive. It seemed to Voûte as if the First Union Astronomer at Johannesburg might "see his way clear to giving me a job and, with the instrument that the Dutch government is lending, carry out cooperative work" with Leiden. The Union Astronomer, Robert Thorburn Ayton Innes, an independent-minded autodidact, might well have looked favorably on Voûte's unusual credentials. Voûte urged De Sitter to propose such an arrangement to Innes. It would be well if De Sitter indicated that the same proposition was going to the Royal Astronomer at the Cape. Johannesburg would be a more congenial site for the Dutch government than Cape Town was, Voûte added, and the letter to Hough at the Cape would sweeten the proposition to Innes: "Between the two institutions reigns no friendship here. A letter to Hough would make Innes accept the plan because it would enable him to steal the honor away from Hough." In the middle of his strategy lesson, Voûte suddenly asked whether two junior-level observers still worked at Leiden, and in case they were leaving whether he could be appointed in their place. He then jumped to discuss the camera mounting that he envisaged for the cooperative South African arrangement.[183] His letter would hardly have been the way to write to a potential patron and employer. Voûte was plainly distraught. He was so uncertain that he wrote to investigate working in the East Indies.

In June 1919, Voûte received an offer from Willem van Bemmelen, director of the Royal Magnetical and Meteorological Observatory at Batavia. Van Bemmelen would allow him to carry out photographical work with the 7-inch refractor, Voûte wrote to De Sitter. Voûte was inclined to accept the post. Johannesburg had fallen through. Although Innes would willingly borrow instruments from the Leiden astronomers, he was in a position to hire "only South Africans ... who fought for their

181. LS De Sitter to Voûte, 5 Mar 1919.
182. LS Voûte to De Sitter, 24 Mar 1919.
183. LS Voûte to De Sitter, 21 Apr 1919.

Empire." In the middle of July, Voûte steamed off for Java. He went without certain knowledge that he would be allowed to devote all his time to astronomy, or even that he would have the use of Leiden's 7-inch refractor. He would be open to new research possibilities, he emphasized to De Sitter. He was ready to return home and begin another chapter in his life.[184] On Java, Voûte became the first director of the new Bosscha Observatory at Lembang.

The way having been prepared, Leiden astronomers took the next, natural step. In 1923 De Sitter arrived at a bilateral exchange agreement with Innes, whom he managed to reward with a doctorate *honoris causa*. The facilities at Johannesburg and at Leiden were thrown open to astronomers from both institutions. Ejnar Hertzsprung, the associate director of the Leiden observatory, travelled to Johannesburg to spend a half year with Innes's new Franklin-Adams 26-cm-aperture photographic telescope. His subject was variable stars. Other observers followed. The arrangement provided first-rate Leiden astronomers with the opportunity of integrating their path-breaking theoretical work with observations of the relatively unknown southern heavens.[185]

De Sitter and Hertzsprung, however, wanted a bigger telescope and complete control over its program, just as Columbia and Yale universities had set up their own, 66-cm refractor at Johannesburg in 1925. The Leiden astronomers turned to the principal international Maecenas of the interwar period, the Rockefeller Foundation. Beginning in 1927, they corresponded with the Foundation's head Wickliffe Rose, and with two Foundation janissaries, physicist Augustus Trowbridge and Princeton chemist Lauder William Jones.[186] In 1930 their ship came in. The Foundation granted 110,000 dollars for the construction and operation of a double-astrograph refracting telescope to be manufactured by Grubb, Parsons & Co. in Newcastle, England, and the government of South Africa expressed its interest in having the telescope at the Union Observatory. Before the money arrived, the Leiden astronomers established the Leiden Observatory Fund to attract donations and funnel

184. LS Voûte to De Sitter, 20 Jun 1919 and 17 Jul 1919. The second letter was penned onstationery belonging to the land-surveying company of Smuts and Shaw.

185. W. Bijleveld, *De Leidse Sterrewacht: Vier Eeuwen wacht bij dag en bij nacht* (Zwolle, 1983) 73-4. *DSB*, "Innis", "De Sitter", and "Hertzsprung", s.v.

186. Leiden, Huygens Laboratorium (in care of Professor C.J. van Houten), Archives of the Leiden Observatory in South Africa [henceforth SA]. Warren Weaver to De Sitter, 19 Jun 1934, where the history of the project is cited.

the observatory's public and private assets into one account. The Rockefeller Foundation was the main but by no means the only contributor (the university provided 800 guilders per year). Nowhere was South Africa mentioned in the Foundation grant or in the establishment of the Observatory Fund, but toward that end the Fund's monies were nearly exclusively spent.[187]

Payments went out to Newcastle for the telescope during the onset of the depression. Leiden asked the Rockefeller Foundation for supplemental increases to take account of unfavorable exchange rates and cuts in funding from the Dutch government, and Warren Weaver authorized the new payments.[188] The Dutch astronomers travelled to Newcastle in 1934 to inspect the objectives; by 1936 they voiced their displeasure with the product. The lenses absorbed much more ultraviolet light than would be suitable for their plans. They proposed to accept the English product at a discount and place a new order with someone else.[189] The Rockefeller Foundation balked at providing between 2000 and 3000 pounds sterling to order the lenses a second time from Grubb. Physicist and assistant director of the Foundation's European office Wilbur Earle Tisdale wrote to Jan Hendrik Oort, associate professor of astronomy at Leiden, that "inasmuch as during the last several years our activities have been strictly limited to fields of biological interest," no more money would be forthcoming.[190]

Leiden decided to go with Zeiss in Jena. Oort informed Tisdale that Zeiss would sell the desired objectives at a great reduction — 50,000 gold marks. In addition, Grubb was willing to cancel part of the contract, amounting to 40,000 gold marks. The remaining sum would come from the Leiden Observatory Fund, Oort continued. He asked the Rockefeller

187. Bijleveld, *Leidse Sterrenwacht* 74. C.J. van Houten, "Stichting Leids Sterrewachtfonds 50 Jaar; een overzicht van de activiteiten", mimeographed report (Leiden, 1980?).

188. SA. Adriaan van de Sande Bakhuyzen (a university curator) to European Director of the Rockefeller Foundation, 26 Oct 1933; Weaver to De Sitter, 19 Jun 1934.

189. SA. De Sitter to W.E. Tisdale, 6 Aug 1934; A. van de Sande Bakhuyzen to Tisdale, 10 Jul 1936.

190. SA. Tisdale to Van de Sande Bakhuyzen, 20 Jul 1937. On the evolution of the Rockefeller Foundation's policy direction: R.E. Kohler, "Warren Weaver and the Rockefeller Foundation Program in Molecular Biology: A Case Study in the Management of Science" in: N. Reingold ed., *The Sciences in the American Context: New Perspectives* (Washington, D.C., 1979) 249-93.

Foundation to pay the 40,000 gold marks to Zeiss instead of to Grubb; two installments were desired, 25,000 marks then and 15,000 to follow.[191] Before receiving a definitive reply from the Foundation, Hertzsprung (who had become director of the Leiden observatory upon De Sitter's death) decided to rent the substandard lenses from Grubb and personally supervise their installation in the buildings that were waiting at Johannesburg.[192] The arrangement would provide a temporary instrument for Hendrik van Gent, the Leiden astronomer who had been assigned to Johannesburg since 1933. In the spring of 1939 the Foundation released a sum to the Leiden Observatory Fund and agreed to hold the final payment of 1301 guilders for Zeiss.[193] But Zeiss miscast the first blanks. A second casting succeeded eighteen months later. By then the war had closed in.[194] Early in 1941 Oort begged Tisdale to convert the unspent balance into dollars, and presumably allow the Leiden astronomers to draw on it from a New York account; Tisdale replied that the funds had been blocked.[195]

Van Gent worked at the Leiden station and at two other observatories in Johannesburg until his death in 1947. The directorship of the Leiden station went to Willem de Sitter's son Aernout, who in 1941 succeeded Voûte at Lembang. Leiden astronomy in South Africa was represented by Willem Hendrik van den Bos, who had been first assistant at the Union Observatory since the late 1920s and who, in 1941, rose to become director. As for the Zeiss lenses, they survived the war in the neighborhood of Jena. In 1949, the Netherlands negotiated their release from the German Democratic Republic and installed them in Johannesburg.[196]

191. SA. Oort to Tisdale, 1 Dec 1937.

192. SA. A. van de Sande Bakhuyzen to Tisdale, 14 Apr 1939.

193. SA. R. Letort (of the Rockefeller Foundation) to A. van de Sande Bakhuyzen, 18 Apr 1939.

194. Bijleveld, *Leidse Sterrewacht* 74-5.

195. SA. Oort to Tisdale, 14 Feb 1941; Rockefeller Foundation to Oort, 20 Mar 1941.

196. J.H. Oort, "In Memoriam Dr. H. van Gent", *Hemel en dampkring* 45 (1947) 159-60; W.S. Finsen, "Obituary: Willem Hendrik van den Bos", *Monthly Notes of the Astronomical Society of Southern Africa* 32 (1958) 1-30; J.H. Oort, "In Memoriam Dr. A. de Sitter", *Hemel en dampkring* 44 (1946) 33-4. Bijleveld, *Leidse Sterrewacht* 75; Van Houten, "Stichting Leids Sterrewachtfonds 50 Jaar".

6. Cultural Imperialism in Three Dimensions

The preceding stories suggest the extent to which the exact sciences have been used to further the overseas political goals of three European nations —France, Germany, and the Netherlands — in their colonies and spheres of influence. Physics and astronomy have facilitated a probe into the means and ends of what might loosely be called cultural imperialism. An open-ended use of 'cultural imperialism' is deliberate. As in an earlier work of mine, here it seems best to refrain from confining the term to an overly restrictive definition: The text supplies the senses in which it has been used.[197] Attitudes toward cultural imperialism exhibit variation in both metropolitan powers and colonial dependencies, but there is a consistent belief that the prosecution of scientific research *in partibus* constituted a beachhead in the overseas entrenchment of metropolitan domination. As one colonial geophysicist put it in his visionary program of 1930, *savants* in the colonies were "agents" for the propagation of metropolitan (in his case, French) "culture," just as "colonists and functionaries are agents of social and economic expansion."[198]

The comparative nature of the preceding discussion permits extraction of preliminary generalizations in the form of a descriptive model. The model centers on three, orthogonal axes of imperialist strategy: the *functionary* axis, the *research* axis, and the *mercantilist* axis. These axes may be thought of as being much like the x-y-z coordinates of analytical geometry. They are suggested as a way of picturing in concrete terms the rather abstract strategies that are under discussion. It lies far from the intent here, however, to enlist the model in the service of 'quantifying' cultural imperialism.[199]

A strategy in the direction of the functionary axis would emulate French practice: a tight union of academic, military, and religious interests where the desire to prosecute original research was depressed, if not extinguished, and where a scientist in foreign parts remained entirely subordinate to metropolitan directives. In the French experience, a

197. Pyenson, *Cultural Imperialism and Exact Sciences.*

198. De Martonne, *Le Savant colonial* (Paris, 1930) 32.

199. Comparative history need not end in typologies. Aristotelian kinds seem difficult to extract from certain cases, such as those in the history of colonial cities. G.A. de Bruijne, "The Colonial City and the Post-Colonial World" in: R. Ross and G.J. Telkamp eds., *Colonial Cities: Essays on Urbanism in a Colonial Context* (Dordrecht, 1985) 231-43, on p. 236.

physicist or astronomer overseas was first and foremost a functionary. He could be recalled if he incurred the displeasure of colonial or metropolitan overlords — be these administrators in a governor general's office, junior ministers in Paris, or Jesuit superiors in France. The guiding lines of late nineteenth- and twentieth-century French policy may be seen succinctly in an address of 1936 by Alfred Lacroix, perpetual secretary of the Academy of Sciences, on scientific research in French overseas territories. In Lacroix's view, there was "no room in our colonies for new, major agencies of purely scientific research." Rather, it was "necessary to locate in the metropolis the center of action for scientific research carried out in *La France d'Outre-Mer*." The metropolitan *savants* would assign younger scientists clearly defined and limited tasks for overseas execution. These tasks accomplished, "the study of the material [acquired] and the discussion of the collected observations would be done in France."[200]

The research axis suggests the drive behind German practice: a loose union of academic, business, and military interests, where the research ethic remained paramount. A German scientist overseas was above all a seeker of new knowledge. He was for the most part on his own, and he prosecuted science by wits and wiles. The ideology may be divined in a proposal to the Prussian Kultusministerium of 1908 written by six distinguished Göttingen scientists who argued in favor of constructing an astrophysical branch of their university's astronomical observatory at Windhoek in German Southwest Africa:

> "Scientific activity in the colonies is certainly the most noble claim that, in this sense, can be advanced. A colonization like that of an observatory already provides a seat of civilization. If it is known that an astrophysical observatory is located at Windhoek, the impression of wasteland and inefficiency, which seems to exist in wider circles about life in the Southwest, may be diminished."[201]

The scientists expressed the view that the prosecution of pure learning would as a matter of course be relevant to the imperial designs of the state. Their proposal, in this case, bore no fruit.

200. A. Lacroix, "Pour une organisation des recherches scientifiques dans nos territoires d'outre-mer", *Actes et comptes-rendus de l'Association Colonies-Sciences* 13 (1937) 21-5, quotations on pp. 23-4. Lacroix's vision, unlike de Martonne's, reflects political realities.

201. Eduard Riecke, Felix Klein, Karl Schwarzschild, Woldemar Voigt, Emil Wiechert, and Hermann Wagner to Kultusministerium, 14 Jul 1908, cited in Pyenson, *Cultural Imperialism and Exact Sciences* 303.

Finally, a strategy in the direction of the mercantilist axis would have scientists serving business interests; research would be subordinated to solving one or another problem in technology. Highly skilled scientists travelled overseas to pioneer Belgian mines in Katanga and set up Canadian nuclear facilities in Argentina, for example, but Belgian and Canadian *savants* did not set down, initially, institutions of higher learning. Belgium and Canada have traditionally favored mercantilist strategies of cultural imperialism.[202]

Where does the Netherlands fit into this picture? Dutch practice involved a union of academic and commercial interests, one where scientists served as official representatives of their state. The state, in fact, generally acted to finance higher learning only when the path had been cleared by private interests and when academic researchers had already committed themselves to the enterprise.[203] A striking expression of the ideology behind this union may be found in the speech given in 1920 by K.A.R. Bosscha, one of Java's richest plantation owners, during the opening ceremonies of the physics laboratory at the Bandung Institute of Technology. Bosscha, who had endowed the building, left the laboratory's research program entirely up to the new professor of physics. The governor general, in his speech, expected technicians to be produced at the institution but did not point to any direction for work at the laboratory. The professor of physics, Jacob Clay, left no doubt that he would be prosecuting pure learning.[204] The Dutch experience, then, would be represented by a vector with positive components projected along all three axes.

In the absence of detailed studies, one may hazard to place the strategies

202. This observation is by no means to deny the use of metropolitan and international scientific organizations, on the part of Canada and Belgium, to further colonial or imperialist ambitions. A striking case of the latter is how King Leopold II created and patronized the International Association for the Exploration and Civilization of Africa in 1876. In 1884-85 the International Association received widespread recognition as the Congo Free State, a *de jure* sovereign power but a *de facto* satrapy of the Belgian monarchy. The charade ended in 1908 when the Belgian parliament assumed control of the Congo Free State as Belgium's first colony. Science was absent from the International Association's agenda, unless the adventures of Sir Henry Morton Stanley, its hired agent, are considered to belong to the annals of geography.

203. Even funding for the Royal Magnetical and Meteorological Observatory at Batavia, in the 1870s, seems to follow this pattern. De Bruin and Pyenson, "Gentleman-Scientist".

204. All the speeches are printed in *Opening van het Natuurkundig Laboratorium.*

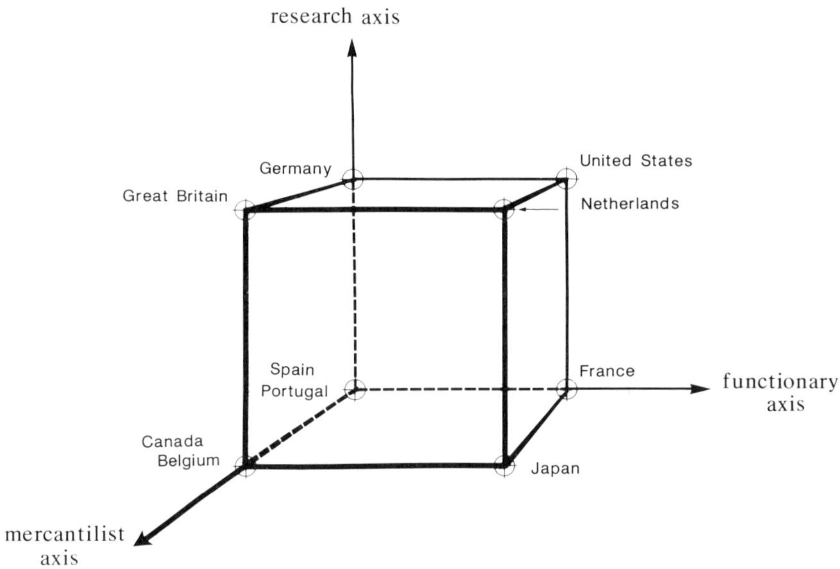

figure 1: Three strategic axes of cultural imperialism.

of other imperialist nations in relation to these three axes. American strategy would rank high in the research ethic and high in responsiveness to metropolitan direction (observatories in Chile, Peru, and Argentina which were controlled by metropolitan astronomers; Rockefeller Foundation investments in China; religious missionaries in Lebanon), but it does not seem centrally concerned with extracting natural resources or selling American goods; United Fruit and Standard Oil, having the United States marines at their command, did not seem to have required the services of scientists as cultural imperialists. British strategy (for the moment leaving aside the educational vortex of the metropolis), while ranking high on the rescarch axis, involved few controls exerted by metropolitan authorities: Canadian, Australian, South African, and Indian physicists followed their karma in carrying out research. For the British, however, unlike for the Americans, the trade interest was paramount: Observatories were erected at coaling stations around the world in order to provide mariners with accurate time signals and weather reports. One may guess that Japanese strategy manifested strong mercantilist and functionary characteristics, while being rather weak along the research axis; at least such seems to have been the pattern in

Japanese management of the Tsingtau observatory, captured from the Germans in 1914. Finally, there will be found imperialist countries exhibiting weak or nonexistent magnitudes in the direction of any of the three axes: Those on the Iberian peninsula come to mind.

These observations have been inserted as points on the accompanying diagram. The reader is cautioned to extend large uncertainty bars around each point — especially those that have not been discussed in the preceding text — and to recall that the evidence relates for the most part to activity in the exact sciences. There is, nevertheless, a certain seductiveness about the display: The major imperialist powers of the twentieth century separate into discrete vectors, and the various national styles of cultural imperialism cohere with more general educational and socio-economic practices.

Whatever fate befalls the model, I think that it suggests how historians must take control of their own generalizations. Theorists in sociology or linguistics may provide inspiration for historical understanding; in this they are like political agitators and street musicians. Historians, having the benefit of a long and broad view, reflect on what has been done before and then invent explanations to fit their documents.

University of Montreal

COMMENTARY

H.W. von der Dunk *On the paper of Lewis Pyenson*

Professor Pyenson's paper on the expansion of Western Science is a rather new and fascinating contribution to the extensive historiography of imperialism. He has limited his research to the exact sciences and has given us a substantial amount of detailed information about institutions, scientific activities and scientists in some of the former colonies, especially those of France, the Netherlands and Germany. He has also put before us some general questions. His exposition can be split into a pure descriptive part and a more general comparative part with some conclusions about differences and similarities on scientific expansion in various European colonies. I shall concentrate on this last point.

Professor Pyenson has first posed the important question: How far or to what extent can purely scientific discourse be isolated from socio-economic reality? He has drawn some conclusions by comparing the French, the German and the Dutch methods of scientific expansion in their colonies.

In the French case a scientist in the colonial or semicolonial world (if for a moment we may put this label on Latin America) was first and foremost a functionary. He was very dependent on his colonial bosses and on the metropolitan authorities. French scientific enterprise had always the function to contribute to the 'grandeur' of France and to promote its political and cultural influence.

In the German case purely scientific interests were dominant. Military, political and commercial interests were certainly not absent, but there was only a loose connection. German scientists felt more free than their French colleagues to concentrate on pure research. Individual and private initiative played an important role. German influence rested first of all on scientific quality, which in several cases was superior. Perhaps it is not too bold to conclude that this was due to the fact that scientists were more free and the supervision of the metropolitan authorities was not so strong.

In the Dutch case in Indonesia — the former Dutch East Indies — we observe a combination of academic and commercial interests. Dutch government came only into action when the path had been cleared by private initiatives and academic research had already started. I think this last conclusion does sound very familiar to Dutch ears of today. There hasn't changed so much. Now all these observations evoke more questions.

First: I wonder if the differences outlined by Pyenson can be explained by socio-economical factors only! Indeed he himself did not introduce them explicitly as far as I can see. The export and expansion of science can be reduced to two main factors: pure scientific interest, which however seems absent as far as sciences such as physics are concerned, and the national interest to promote national culture and to export Western civilization. Germany is perhaps an example. Pyenson has demonstrated that the political and military arguments were second-hand factors. I think this has partly to do with the traditional continental-European orientation of German politics in the 19th century. The political and military elites in Germany, especially in Prussia, did not have a specific inclination in overseas territories. Imperialism and colonialism only started in the eighties and nineties when the new industrial elite backed an imperial policy. This may partly explain the loose connection between scientific, military and commercial interests. After World War I, when Germany lost its colonies and its political and commercial ties with the countries oversea were weakened or even broken by the treaty of Versailles, scientific export became the only instrument to regain entry to the lost territories.

Besides, in Germany there existed a kind of silent deal between the state and the cultural or academic elite: politics and science were separated, although or perhaps because the intellectual elite, e.g. the academicians, were servants of the State. In exchange for their political abstinence or obedience, the State granted them freedom in their own field. So the connection between culture and politics was quite different compared to France.

In France at the one hand the intellectuals since Voltaire had a long tradition of critical political engagement and were used to play this role in society. On the other hand the entire educational system was strongly centralized. So intellectuals and scientists could identify with the national interest and with the 'grandeur' of France, which was not their master but something like their other self. Another element seems to me the steady decline of France after 1871. So imperialism and cultural export became a substitute for the loss of influence and power in Europe. France had to concentrate on the non-European regions to regain its old proud and dominant position and to satisfy its longing for 'grandeur'. I think the examples of French activities in Canada show that this need to oppose the influence of other European powers was an important motive for scientific imperialism. The much older and greater colonial and maritime tradition of France, compared to Germany, can also explain the much more political (or functionary) impact of French science export.

Now it would be most instructive to have a more detailed additional survey of English and American cultural imperialism. Why did the Americans combine the scientific with the functionary motive and the English with the commercial interest? What was the relation to the political background? We cannot separate all these activities — the French, German and Dutch as well — from the situation in the metropolitan countries, not only the political or economical situation, but also the career perspectives and the conditions at the universities and other scientific institutions. When I just said that the pure scientific interest was one of the two central motives behind all this, I must add of course that this motive was in many cases mixed with personal ambitions. The absolute idealistic attitude amongst scientists may be as rare as amongst other human beings.

The Iberian peninsula, Professor Pyenson has remarked, exhibited only weak attitudes in relation to these strategies and scientific export. This is without doubt a result from their weak position in Europe and specially from the fact that Spain and Portugal did seriously fall behind in modernization and industrialization and also in scientific development. Both actually did belong during this period to the 'underdeveloped' countries of Europe.

The increased activities after World War I cannot be separated from the increasing consciousness in the European countries that the highdays of imperialism were over. So those nations had to switch to the cultural-intellectual field to strengthen or continue their grasp on the oversea-world. It is striking that every idea of a sort of Western cultural solidarity seemed completely absent. One of the impulses to expand and thus also to export science was the national interest and the inter-European rivalry. Of course we must be careful not to generalize this thesis and things certainly could differ from case to case.

It would be interesting also to know more about the origins of the functional approach. Can we simply speak of some kind of strategy? Which means a conscious policy, conceived at a certain moment by certain authorities or persons? Or was e.g. the French method and way of cultural imperialism primarily a projection of the French system and tradition? As was the Dutch attitude. It would be very important, also in view of the present and future situation in the world, to know more about the results of the scientific impact on the former colonial peoples. What were the differences and can these differences in reactions be connected with differences in method of the metropolitan countries as outlined in

Pyenson's diagram? That the differences are also a result of the different cultures and traditions amongst the former colonial or semi-colonial peoples goes without saying. I realise of course that these enormous questions already belong to another lecture.

University of Utrecht

INDEX OF NAMES

NIEUWE NEDERLANDSE BIJDRAGEN TOT DE GESCHIEDENIS DER GENEESKUNDE EN DER NATUURWETENSCHAPPEN

Editors: M.J. van Lieburg, L.C. Palm,
H.A.M. Snelders and R.P.W. Visser

No. 1.
Lindeboom, G.A.: Descartes and Medicine. Amsterdam 1979. 134 pp.
Hfl. 30,—

No. 2.
Lindeboom, G.A.: Adriaan van den Spiegel (1578-1625) hoogleraar in de ontleed- en heelkunde te Padua. Amsterdam 1978. 125 pp. uitverkocht

No. 3.
Lieburg, M.J. van: Het medisch onderwijs te Rotterdam (1467-1967). Amsterdam 1978. 111 pp. Hfl. 18,—

No. 4.
Snelders, H.A.M.: Het Gezelschap der Hollandsche Scheikundigen — Amsterdamse chemici uit het einde van de achttiende eeuw. Amsterdam 1980. 111 pp. Hfl. 20,—

No. 5.
Brinkman, A.A.A.M.: De alchemist in de prentkunst. Amsterdam 1982. 69 pp. 4 to. Hfl. 25,—

No. 6.
Dumaître, Paule: La curieuse destinée des planches anatomiques de Gérard de Lairesse. Amsterdam 1982. XIV,107 pp. Hfl. 25,—

No. 7.
Deutsch-Niederländischen Beziehungen in der Medizin des 17. Jahrhunderts. Vorträge des Deutsch-Niederländischen Medizinhistorikertreffens 1981. M.J. van Lieburg und R. Toellner (Hrsg.). Amsterdam 1982. 108 pp.
Hfl. 25,—

No. 8.
Antoni van Leeuwenhoek 1632-1723. Studies on the life and work of the Delft scientist commemorating the 350th anniversary of his birthday. Edited by L.C. Palm & H.A.M. Snelders. Amsterdam 1982. 212 pp.
Hfl. 50,—

No. 9.
Berkel, K. van: Isaac Beeckman (1588-1637) en de mechanisering van het wereldbeeld. Isaac Beeckman (1588-1637) and the mechanization of the world picture (with a summary in English). 344 pp. Amsterdam 1983.
Hfl. 50,—

No. 10.
Fahrenheit's Letters to Leibniz and Boerhaave. Edited, translated and annotated by Pieter van der Star. Amsterdam/Leiden 1983. 195 pp.
Hfl. 50,—

No. 11.
Schraders, Catharina Geertruida: Memoryboeck van de Vrouwens. Het notitieboek van een Friese vroedvrouw 1693-1745. Bewerkt en ingeleid door Drs. M.J. van Lieburg. Met een verloskundig commentaar van Prof.Dr. G.J. Kloosterman. Amsterdam 1984. 218 pp. Hfl. 35,—

No. 12.
Visser, Robert Paul Willem: The Zoological Work of Petrus Camper (1722-1789). Het zoölogisch werk van Petrus Camper (1722-1789) (met een samenvatting in het Nederlands). Amsterdam 1985. 207 pp. Hfl. 45,—

No. 13.
Kracht en Stof. De introductie van moderne natuurwetenschappelijke denkwijzen in de geneeskunde, zoals blijkt uit Nederlandse medische vakbladen, 1840-1870. Verslag van een symposium, gehouden te Nijmegen op 21 september 1985, onder redactie van D. de Moulin. Amsterdam 1985. 127 pp. Hfl. 30,—

No. 14.
Deutsch-Niederländischen Beziehungen in der Medizin des 18.Jahrhunderts. Vorträge des Deutsch-Niederländischen Medizinhistorikertreffens 1982. Herausgegeben von R. Toellner und M.J. van Lieburg. Amsterdam 1985. 92 pp. Hfl. 25,—

No. 15.
Napjus, J.W.: De hoogleraren in de geneeskunde aan de hogeschool en het athenaeum te Franeker. Verzamelde opstellen. Ingeleid en voortgezet door Dr. G.A. Lindeboom. Amsterdam 1985. XIX,210,XXVI pp. Gebonden.
Hfl. 40,—

No. 16
Theunissen, Bert: Eugène Dubois en de aapmens van Java. Een bijdrage tot de Geschiedenis van de paleoantropologie. Amsterdam 1985. 302 pp.
Hfl. 50,—

No. 17.
Nineteenth-Century Scientific Instruments and their Makers. Papers presented at the Fourth Scientific Instrument Symposium, Amsterdam 23-26 October 1984. Edited by P.R. de Clercq. Amsterdam/Leiden 1985. 275 pp.
Hfl. 70,—

No. 18.
Lieburg, M.J. van: Het Bataafsch Genootschap der Proefondervindelijke Wijsbegeerte te Rotterdam 1769-1984. Een bibliografisch en documenterend overzicht. Amsterdam 1985. 283 pp. Hfl. 60,—
No. 19.
Hendrik Engel's Alphabetical List of Dutch Zoological Cabinets and Menageries. Second, enlarged edition. Prepared by Pieter Smit with the assistance of A.P.M. Sanders and J.P.F. van der Veer. Amsterdam 1986. X,340 pp. Bound. Hfl. 70,—
No. 20.
Hakfoort, Casper: Optica in de eeuw van Euler. Opvattingen over de natuur van het licht 1700-1795. Optics in the age of Euler. Conceptions of the nature of light, 1700-1795. (with a summary in English). Amsterdam 1986. 219 pp. Hfl. 49,50
No. 21.
Lieburg, M.J. van: Het Coolsingelziekenhuis te Rotterdam (1839-1900). De ontwikkeling van een stedelijk ziekenhuis in de 19e eeuw. Amsterdam 1986. XVIII,770 pp. 4to. Bound. Hfl. 160,—
No. 22.
Mother and Child were Saved. The memoirs (1693-1740) of the Frisian midwife Catharina Schrader. Translated and annotated by Hilary Marland. With introductory essays by M.J. van Lieburg and G.J. Kloosterman. Amsterdam 1987. 89 pp. Hfl. 25,—
No. 23.
Wiggen, G.J. van: In meer eerbare banen. De ontwikkeling van het tandheelkundig beroep in Nederland van 1865-1940. Amsterdam 1987. 369 pp. Hfl. 65.—
No. 24.
Mijnhardt, W.W.: Tot Heil van 't Menschdom. Culturele genootschappen in Nederland, 1750-1815. Amsterdam 1988. 430 pp. Hfl. 75,—
No. 25.
Lieburg, M.J. van: Bevorderlijk voor de kunst en nuttig voor de maatschappij. De geschiedenis van het Genootschap ter bevordering van Heel- en Verloskunde (1857-1929) en van het Amsterdams Geneeskundig Genootschap (1925-1988). Amsterdam 1988. 124 pp. Hfl. 25,—
No. 26.
's-Rijkskweekschool voor militaire geneeskundigen te Utrecht (1822-1865). Verslag van een symposium gehouden te Nijmegen op 26 september 1987 onder redactie van D. de Moulin met een lijst van kwekelingen, samengesteld door M.J. van Lieburg. Amsterdam 1988. 193 pp. Hfl. 45,—

DURHAM UNIVERSITY LIBRARY
3 0104 00649593 5

No. 27.
A.I. Bierman: Van artsenijmengkunde naar artsenijbereidkunde. Ontwikkelingen van de Nederlandse farmacie in de negentiende eeuw. Amsterdam 1988. 312 pp. Hfl. 65,—

No. 28
Pieter Smit: Artis. Amsterdam 1988. 392 pp. 27 ill., 7 kleurenfoto's. Hfl. 45,—

No. 29
M.J. van Lieburg en H.A.M. Snelders: "De bevordering en volmaking der proefondervindelijke wijsbegeerte". De rol van het Bataafsch Genootschap te Rotterdam in de geschiedenis van de natuurwetenschappen, geneeskunde en techniek (1769-1988). Amsterdam 1989. 267 pp. Hfl. 60,—